SWITCH YOUR FRAME

50 Cognitive Shifts for Career, Relationships, and Resilience

Micah Deunald

COPYRIGHT INFORMATION

ISBN:978-1-291-84527-3

TRADEMARK DISCLAIMER

DISCLAIMER

The information provided in this book is for general educational and informational purposes only. It is not intended as a substitute for professional psychological, psychiatric, medical, legal, or financial advice. The author and publisher are not licensed mental health professionals, and the content herein should not be construed as therapy, counseling, or treatment for any mental health condition.

Readers are encouraged to seek the guidance of qualified professionals for specific concerns regarding their mental health, relationships, career, or personal circumstances. The author and publisher disclaim any liability for any adverse effects arising from the use or application of the information contained in this book.

The examples, case studies, and scenarios presented in this book are composites or hypothetical illustrations created for educational purposes. Any resemblance to actual persons, living or dead, or actual events is purely coincidental.

For everyone who has ever felt stuck

in a way of seeing that no longer serves them.

The frame can be switched.

The Teaser

Real change requires more than just intellectual understanding. It requires visceral, physical, and practical application.

You are holding the map, but I want to give you the tools.

To ensure you don't just *read* about switching your frame but actually *do* it, I have created a digital companion for this journey: **The "Switch Your Frame" Secret Vault.**

This hidden collection contains the "raw" materials that couldn't fit on these pages—audio tracks to regulate your nervous system, deep-dive shadow workbooks, and verbatim scripts for your toughest conversations.

Keep reading. When you reach **the last page**, you will find the key to unlock the Vault.

Don't just read this book. Live it.

"It is not things that disturb us,

but our judgments about things."

— Epictetus, Enchiridion

TABLE OF CONTENTS

INTRODUCTION

The Operating System You Never Knew You Had

It is not things that disturb us, but our judgments about things.

— Epictetus, Enchiridion

Two people sit in the same meeting. The boss announces a restructuring. One person thinks: *This is my chance to step up*. The other thinks: *I'm going to lose my job*. Same room. Same words. Same facts. Completely different internal experiences. And here's what matters: completely different actions will follow.

The first person will volunteer for new responsibilities, position themselves for the transition, and stay visible to decision-makers. The second will withdraw, update their resume in a panic, and spend the next three weeks catastrophizing instead of contributing. Six months later, one got promoted. The other got managed out—not because of the restructuring, but because of everything that happened after.

This is not a story about positive thinking. This is a story about the invisible architecture that shapes every outcome in your life.

You have an operating system running beneath your conscious awareness. It takes in raw data—an email, a silence, a raised eyebrow, a quarterly report—and instantly converts it into meaning. That meaning determines how you feel. How you feel influences what you do. What you do compounds into your results.

Most people never examine this operating system. They assume their interpretations are reality. When they feel anxious, they believe something anxiety-worthy is happening. When they feel angry, they believe someone did something worth being angry about. They live their entire lives reacting to their own mental constructions as if those constructions were facts.

This book is about something different. It's about gaining access to the source code.

• • •

Nearly two thousand years ago, a former slave named Epictetus taught philosophy in a small school in Greece. His opening lesson—the very first thing he wanted his students to understand—was this: we are disturbed not by events themselves, but by our judgments about events.

This wasn't abstract philosophy. Epictetus had been enslaved, beaten, and walked with a permanent limp from abuse. He knew suffering. And he knew that the same circumstances could break one person and strengthen another. The variable wasn't the circumstance. It was the interpretation.

Marcus Aurelius, a Roman emperor who could have had anything, wrote in his private journal: "You have power over your mind—not outside events. Realize this, and you will find strength." He wasn't being idealistic. He was running an empire during plagues, wars, and constant political threats. He needed tools that worked.

For centuries, this insight remained philosophical. Then in the 1960s, a psychiatrist named Aaron Beck noticed something strange. His depressed patients weren't just sad—they were thinking in systematically distorted ways. They were catastrophizing, overgeneralizing, and filtering out positive information. Change their thinking patterns, he discovered, and their depression lifted.

Beck developed Cognitive Behavioral Therapy based on this principle. The research followed. A meta-analysis examining cognitive restructuring—the formal term for changing how you interpret events—found an effect size of .35 with a Cohen's d of 0.85 on therapeutic outcomes. In plain language: learning to think differently produces measurable, significant improvements in how people function.

Stoic philosophy and modern psychology arrived at the same destination from opposite directions. The ancient insight and the clinical evidence now reinforce each other. The way you frame events changes everything that follows.

• • •

Before we go further, let me be clear about what this book is not.

This is not a book about positive thinking. Positive thinking tells you to replace negative thoughts with positive ones. Tell yourself you're great. Visualize success. Believe and you will achieve. The problem is that your brain isn't stupid. When you try to convince yourself that something bad is actually good, your mind rebels. It knows you're lying. The anxiety doesn't decrease—it goes underground and comes back stronger.

Cognitive restructuring, the approach this book teaches, works differently. It doesn't ask you to pretend things are good when they're not. It asks you to examine whether your interpretation is accurate, complete, and useful.

Consider the difference. Positive thinking says: *I didn't get the promotion, but that's great because something better is coming!* Your brain responds: *That's delusional. I wanted that promotion. This hurts.*

Cognitive restructuring says: *I didn't get the promotion. That's disappointing. But does this single event mean I'll never advance? Does it mean I'm not valued? Does it mean my career is over? What's the actual evidence here?*

One approach denies reality. The other interrogates your assumptions about reality. The Stoics called this "testing your impressions." Modern therapists call it "examining the evidence." Both are doing the same thing: putting space between what happened and what you've decided it means.

This matters because most of your suffering doesn't come from events. It comes from the stories you tell yourself about events. You didn't just miss the promotion—you're "falling behind." Your partner didn't just seem distant today—they're "pulling away." The project didn't just hit a setback—it's "falling apart."

These stories feel like facts. They're not. They're interpretations layered on top of facts. And once you see the difference, you have options you didn't have before.

• • •

I want to be honest about what's ahead.

This book will give you fifty cognitive shifts—specific reframes you can apply to common situations in your career, relationships, and moments of adversity. Each shift moves you from a reactive, limiting interpretation to a more accurate, useful one. Over time, these shifts compound. You'll notice yourself catching distorted thinking faster. You'll have language for what you're experiencing. You'll respond instead of react.

What this book won't do is eliminate difficulty from your life. Cognitive restructuring doesn't make hard things easy. It makes hard things manageable. You'll still feel disappointment, frustration, and anxiety. The goal isn't emotional numbness. The goal is emotional accuracy—feeling what's proportionate to what's actually happening, rather than to what your catastrophizing mind imagines is happening.

This book also won't work through reading alone. The shifts have to be practiced. You have to catch yourself mid-thought, recognize the distortion, and apply the reframe. This is a skill. Like any skill, it's awkward at first. You'll forget. You'll slip back into old patterns. Then you'll remember, try again, and gradually get better.

The payoff is this: you stop being at the mercy of your automatic interpretations. You start choosing your responses. Viktor Frankl, the psychiatrist who survived Nazi concentration camps, wrote: "Between stimulus and response there is a space. In that space is our power to choose our response. In our response lies our growth and our freedom."

This book is about expanding that space.

• • •

The fifty shifts in this book are organized into three domains: Career, Relationships, and Resilience.

Career covers how you interpret professional setbacks, criticism, uncertainty, comparison, and the meaning of your work. These are the situations where cognitive distortions cost you opportunities—where catastrophizing makes you withdraw, where personalization makes you defensive, where all-or-nothing thinking makes you quit too early or hold on too long.

Relationships covers how you interpret conflict, disappointment, other people's behavior, and your own patterns. These are the situations where distorted thinking erodes connection—where mind-reading creates distance, where the fairness fallacy breeds resentment, where control fantasies make you try to change people who can only change themselves.

Resilience covers how you interpret adversity, strong emotions, and the passage of time. These are the situations where your mental framing determines whether difficulty breaks you or builds you—where the story you tell about hardship becomes the hardship itself.

Each domain builds on the previous. The foundational shifts in Part One apply everywhere. The career shifts teach you to manage external pressures. The relationship shifts teach you to manage interpersonal complexity. The resilience shifts teach you to manage your own internal weather. Together, they form something like an operating system upgrade—a coherent set of mental frameworks for navigating life.

• • •

Each chapter follows the same structure: Theory, Shift, Action.

Theory explains the psychological principle at work. Why do we distort our thinking in this particular way? What cognitive error is operating? What does the research show about how this pattern develops and how it can change? This isn't academic padding—it's the foundation that makes the shift stick. When you understand why your mind defaults to a certain interpretation, you gain leverage over that default.

Shift presents the specific reframe. Each shift has a "from" and a "to"—the limiting interpretation you're moving away from and the more accurate interpretation you're moving toward. These are designed to be memorable and applicable. When you find yourself in the relevant situation, the shift gives you a concrete alternative to your automatic response.

Action provides the practical exercises. Reflection questions to identify your patterns. Implementation steps to apply the shift in real situations. Troubleshooting for common obstacles. This is where the work happens. Reading about cognitive restructuring is not the same as doing cognitive restructuring. The action sections are how you close the gap.

You can read this book straight through or use it as a reference. If you're struggling with a specific situation—a difficult boss, a strained relationship, a setback you can't shake—find the relevant chapter and work through it. If you want the full framework, start at the beginning and build the foundation first.

Either way, don't just read. Do the exercises. Try the shifts. Notice what happens. Adjust and try again. This is a practice, not a concept.

• • •

Here's what I'm promising you:

Right now, you feel reactive. Things happen to you and you respond automatically. Your boss sends a terse email and your stomach clenches. Your partner makes a comment and you're suddenly in an argument. A project goes sideways and you spiral for days. You're at the mercy of events—and worse, you're at the mercy of your own interpretations of events.

If you work through this book, you will feel different. Not immune to difficulty—that's not the goal. But strategic about your responses. Calm in situations that used to hijack you. Capable of distinguishing between what actually happened and what you've decided it means.

You will move from reactive to intentional. From stuck to in motion. From feeling like life happens to you to feeling like you're the architect of your experience.

This isn't about becoming emotionless or detached. It's about becoming accurate. Feeling what's proportionate to what's real, not what your worst-case-scenario thinking has manufactured. Acting on facts, not stories. Responding to situations as they are, not as your anxious mind insists they must be.

The Stoics had a term for this: they called it *prohairesis*—the faculty of choice. They believed it was the only thing truly under our control. Not outcomes, not other people, not circumstances. Just this: the ability to choose how we interpret and respond to whatever happens.

Modern psychology calls it psychological flexibility. The research shows it correlates with life satisfaction, flourishing, and emotional balance. The Stoics and the scientists are describing the same thing.

You already have this faculty. You've just never been shown how to use it deliberately.

That's what the next fifteen chapters are for.

Let's begin.

PART ONE:

THE FOUNDATION

CHAPTER 1: The Lens You Don't Know You're Wearing

The invisible architecture of interpretation

> *We see the world not as it is, but as we are.*
>
> — Anaïs Nin

Your partner comes home and walks past you without saying hello. What just happened?

Option A: They're angry at you. You must have done something wrong. The relationship is in trouble. You need to figure out what you did and fix it immediately.

Option B: They had a terrible day at work. They're preoccupied and didn't even register that they walked past you. It has nothing to do with you.

Option C: They're feeling unwell. They're heading straight to bed and will explain later.

Here's the thing: you don't know which one is true. You cannot know from the available evidence. But your brain picked one anyway. It picked one instantly, automatically, and with complete confidence. And whatever it picked determined what you felt next—and what you did next.

If your brain picked Option A, your stomach tightened. You started rehearsing conversations in your head. You became defensive before any accusation was made. When your partner emerged from the bedroom, you were already primed for conflict.

If your brain picked Option B or C, you felt mild curiosity. Maybe some concern for them. When they emerged, you asked how their day was. The evening unfolded completely differently.

Same event. Same silence. Same walk past. Radically different experiences—all determined by an interpretation you didn't consciously choose.

This is the lens you don't know you're wearing.

• • •

The Event-Interpretation-Response Chain

In the 1960s, psychiatrist Aaron Beck made a discovery that would reshape how we understand psychological distress. He noticed that his depressed patients weren't just experiencing sadness—they were interpreting events in systematically distorted ways. A neutral comment became evidence of rejection. A minor setback became proof of worthlessness. The pattern was so consistent that Beck could predict how a depressed patient would interpret almost any situation.

This led to a framework that underlies all of cognitive behavioral therapy: the Event-Interpretation-Response chain.

It works like this:

Event: Something happens. Your boss sends a short email. Your friend cancels plans. You make a mistake at work. These are facts—observable, verifiable occurrences in the world.

Interpretation: Your mind assigns meaning to the event. The short email means your boss is upset with you. The cancelled plans mean your friend doesn't value you. The mistake means you're incompetent. These are not facts—they're stories your mind generates about facts.

Response: Based on your interpretation, you feel something and do something. If the short email means your boss is upset, you feel anxious and spend the afternoon catastrophizing instead of working. If it just means your boss was busy, you feel nothing and move on.

The critical insight is this: your response follows from your interpretation, not from the event itself. Two people can experience the identical event and have completely different responses—because they interpreted it differently.

Most people collapse the first two steps. They experience Event → Response as if interpretation weren't involved. The email arrives, they feel anxious. The plans get cancelled, they feel rejected. It seems like the event caused the feeling directly.

But there's always an interpretation in between. Always. The question is whether you're aware of it.

• • •

The Difference Between Facts and Stories

A fact is what happened. A story is what you decided it means.

Fact: Your colleague didn't include you in the meeting invite.

Story: They're trying to exclude you. They don't respect your input. You're being pushed out of the project.

Alternative story: They forgot. It was an oversight. The meeting isn't relevant to your role. They assumed you were too busy.

The fact supports both stories equally well. You cannot determine which is true from the fact alone. But your brain doesn't wait for more information. It picks a story—usually the threatening one—and treats it as truth.

Here's a simple test to distinguish facts from stories: Could a video camera record it?

A camera could record that the email was three sentences long. It cannot record that the email was "cold" or "dismissive." Those are interpretations.

A camera could record that your partner walked past without speaking. It cannot record that they were "ignoring you" or "being passive-aggressive." Those are stories you added.

A camera could record that you received critical feedback. It cannot record that you "failed" or that your boss "doesn't believe in you." Those are meanings you assigned.

The gap between what the camera recorded and what you experienced—that gap is where all of your unnecessary suffering lives.

• • •

Why Your Brain Defaults to Threat Detection

If your brain systematically picks the worst interpretation, that's not a malfunction. It's a feature.

For most of human history, the cost of missing a threat was death. The cost of seeing a threat that wasn't there was a moment of unnecessary anxiety. Evolution optimized for survival, not accuracy. Your ancestors who assumed the rustling in the bushes was a predator lived to reproduce. The ones who assumed it was just the wind sometimes didn't.

This is called the negativity bias. Your brain is wired to detect, remember, and prioritize negative information over positive information. It takes roughly five positive interactions to offset one negative interaction in a relationship. Bad news travels faster than good news because we're built to attend to it.

In the modern world, this creates a problem. Most of the "threats" you perceive aren't threats to your survival. A terse email from your boss will not kill you. Being excluded from a meeting will not end your career. Your partner's bad mood is not evidence of impending abandonment.

But your brain responds as if they might be. It generates the same alarm signals, the same cortisol spike, the same defensive posture. You're using threat-detection hardware built for the savanna to navigate office politics and relationship dynamics.

Understanding this doesn't eliminate the bias, but it gives you leverage. When you notice yourself jumping to the worst interpretation, you can ask: Is this a genuine threat, or is this my ancient alarm system misfiring?

The answer is almost always the latter.

• • •

How Cognitive Distortions Compound

A single distorted interpretation doesn't cause much damage. The problem is that distortions compound.

Here's how it works. You interpret an ambiguous event negatively. That interpretation generates anxiety. The anxiety primes you to interpret the next ambiguous event negatively. That generates more anxiety. The pattern accelerates. Soon you're seeing threats everywhere, your baseline anxiety is elevated, and your interpretations become increasingly disconnected from reality.

Psychologists have identified specific patterns of distorted thinking that show up repeatedly. These are called cognitive distortions, and they function like mental shortcuts that consistently lead you wrong.

Catastrophizing: Jumping to the worst possible outcome. "I made a mistake in the presentation; I'm going to get fired."

Mind-reading: Assuming you know what others are thinking without evidence. "She didn't smile at me; she must hate me."

All-or-nothing thinking: Seeing things in black-and-white categories. "If I don't do this perfectly, I've failed completely."

Personalization: Assuming events are about you when they're not. "The team lost the client because of me."

Overgeneralization: Drawing broad conclusions from single events. "I failed this interview; I'll never get a good job."

These distortions don't operate in isolation. They reinforce each other. You personalize an event ("This is my fault"), then catastrophize the consequences ("I'm going to be fired"), then overgeneralize ("I always ruin everything"), then mind-read the people around you ("Everyone can see I'm a failure").

The spiral builds quickly. And it builds on a foundation of interpretations, not facts.

• • •

The Three Foundational Shifts

Before we address specific situations in career, relationships, and resilience, you need three foundational shifts. These are the base layer that everything else builds on. Get these right, and the specific shifts in later chapters become much easier to apply.

SHIFT 1

FROM: *"This is reality."*

TO: *"This is my interpretation of reality."*

This is the most fundamental shift in the book. Everything else depends on it.

When you think "My boss is upset with me," you're not reporting a fact. You're reporting an interpretation of limited data—an email, a tone of voice, a facial expression. The interpretation feels like reality because your brain presents it that way. But it's not reality. It's your mind's best guess based on incomplete information filtered through your biases, fears, and past experiences.

The shift is subtle but powerful. Instead of "My boss is upset with me," you think "I'm interpreting this as my boss being upset with me." The content is the same. But the framing changes everything. Now you're not dealing with a fact—you're dealing with a hypothesis. And hypotheses can be examined, tested, and revised.

This doesn't mean your interpretation is wrong. Sometimes it's right. The point is that you can't know it's right until you examine it. And you can't examine it while you're treating it as an established fact.

SHIFT 2

FROM: *"I know what this means."*

TO: *"I'm making an assumption about what this means."*

This shift targets the confidence problem. Your brain doesn't present interpretations tentatively. It presents them with certainty. You don't think "Maybe she's upset with me"—you think "She's upset with me." The certainty is an illusion.

When your colleague doesn't return your message, you "know" they're ignoring you. When your partner seems quiet, you "know" something is wrong. When your friend makes a comment, you "know" what they meant by it.

You don't know any of these things. You're guessing. Your guesses are informed by past experiences, pattern recognition, and cognitive shortcuts. Some of your guesses are probably right. Many are wrong. But you're treating all of them as certain knowledge.

The shift is to add the word "assuming" back into your thinking. "I'm assuming she's upset with me." "I'm assuming he's ignoring my message." "I'm assuming that comment was a criticism."

This opens space for alternatives. If you're assuming, then other assumptions are possible. Maybe she's busy. Maybe he hasn't seen your message. Maybe the comment was neutral and you added the critical tone yourself.

SHIFT 3

FROM: *"They made me feel this way."*

TO: *"I chose to interpret it this way."*

This is the ownership shift. It's also the most uncomfortable.

We like to believe that other people cause our feelings. "He made me angry." "She made me feel inadequate." "They made me feel excluded." This framing is convenient because it locates the problem outside ourselves. If they caused the feeling, then they need to change—not us.

But it's not accurate. Events don't cause feelings directly. Events trigger interpretations, and interpretations generate feelings. The feeling of anger arose because you interpreted his actions as disrespectful. The feeling of inadequacy arose because you interpreted her success as evidence of your failure. The feeling of exclusion arose because you interpreted their behavior as rejection.

Different interpretations would have produced different feelings. Someone with a different lens might have seen the same events and felt nothing—or felt something entirely different.

This isn't about blame. It's not your "fault" that you feel what you feel. Your interpretive patterns were shaped by your history, your temperament, and your experiences. But those patterns are still yours. And recognizing that gives you something important: the possibility of change.

If they made you feel this way, you're powerless. If your interpretation made you feel this way, you have options.

• • •

Putting It Together

These three shifts work as a sequence.

Something happens. Your automatic interpretation fires. You notice the interpretation and apply Shift 1: *This is my interpretation, not reality.* Then you apply Shift 2: *I'm assuming this is what it means; other meanings are possible.* Then you apply Shift 3: *My interpretation is creating my feeling; if I interpret differently, I'll feel differently.*

This doesn't happen instantly. At first, you'll catch yourself after the fact—sometimes hours or days later. That's fine. You're building a new skill. With practice, the gap closes. You'll catch distorted interpretations faster. Eventually, you'll notice them in real time and have the option to reframe before you spiral.

The goal isn't to eliminate automatic interpretations. Your brain will keep generating them. The goal is to stop treating them as truth. To create a gap between the interpretation and your response. To remember, in the moment, that you're wearing a lens—and lenses can be adjusted.

TAKE ACTION: YOUR NEXT STEPS

Reflection

Think of a recent situation where you experienced a strong negative emotion—anxiety, anger, hurt, or frustration. Answer these questions:

1. What was the event? Describe only what a video camera would have recorded. Strip away all interpretation. What actually happened?

2. What was your interpretation? What did you decide the event meant? What story did you tell yourself about why it happened and what it said about you, the other person, or your situation?

3. What was your response? What did you feel? What did you do? How did your interpretation shape what came next?

4. What alternative interpretations were possible? List at least two other ways you could have interpreted the same event. What feelings and actions would those interpretations have produced?

Implementation

This week, practice catching interpretations in real time. When you notice a strong emotional reaction, pause and ask yourself: What just happened (the event)? What am I telling myself about it (the interpretation)? Is my interpretation the only possibility?

Keep an Interpretation Log. For the next seven days, write down at least one situation per day using the Event-Interpretation-Response format. Note the event, your automatic interpretation, the feeling it produced, and one alternative interpretation you could have made.

Practice the language of interpretation. When you catch yourself thinking in certainties ("He's upset with me"), rephrase to include the interpretation ("I'm interpreting this as him being upset with me"). The language change creates mental space.

Troubleshooting

"I can't separate the event from the interpretation—they feel like the same thing." This is normal at first. Try asking: What would someone with no emotional investment in this situation say happened? What would a stranger describe? That neutral description is the event. Everything else is interpretation.

"My interpretation is obviously correct. There's no other way to see it." This feeling is a sign that you're fused with your interpretation. Force yourself to generate alternatives anyway, even if they seem unlikely. The exercise isn't about finding the "right" interpretation—it's about breaking the automatic assumption that your first interpretation is the only one.

"I catch the interpretation too late, after I've already reacted." That's fine. Catching it afterward still builds the skill. Over time, the gap will close. What matters is that you're practicing the pattern recognition.

• • •

You now have the foundational framework: events trigger interpretations, and interpretations generate responses. You have three shifts to interrupt the automatic process. And you have the beginning of a practice to build this skill.

In the next chapter, we'll add another layer to the foundation: the dichotomy of control. You'll learn to distinguish between what you can actually change and what you can only accept—and how getting this distinction wrong is the source of most unnecessary suffering.

CHAPTER 2: The Dichotomy That Changes Everything

The ancient framework that modern psychology validates

> *Some things are within our power, while others are not. Within our power are opinion, motivation, desire, aversion, and, in a word, whatever is of our own doing; not within our power are our body, our property, reputation, office, and, in a word, whatever is not of our own doing.*
>
> — Epictetus, Enchiridion

You've just given the presentation of your career. You prepared for weeks. You rehearsed until you could deliver it in your sleep. You answered every question with precision. You left the room feeling like you nailed it.

Now you wait.

The decision isn't yours. It never was. A committee will deliberate. They'll weigh factors you don't know about—internal politics, budget constraints, competing priorities, personal preferences. They might love your proposal and reject it anyway. They might have concerns you never anticipated. They might have made their decision before you walked in the room.

You can spend the next two weeks obsessing over the outcome. Replaying the presentation. Analyzing every facial expression in the room. Checking your email compulsively. Losing sleep. Snapping at your family. Accomplishing nothing at work because your mind is elsewhere.

Or you can recognize a fundamental truth: the outcome was never in your control. Your preparation was. Your delivery was. Your follow-up might be. But the decision itself? That belongs to other people, operating on information and priorities you'll never fully understand.

This distinction—between what you control and what you don't—is the most liberating idea you'll encounter in this book. It's also nearly two thousand years old.

• • •

The Stoic Foundation

Epictetus was born a slave in the Roman Empire around 50 CE. He was owned, beaten, and walked with a permanent limp—possibly from abuse, possibly from a childhood illness. He had every reason to feel powerless. Instead, he developed a philosophy centered on the one thing no one could take from him: his capacity to choose his response to circumstances.

The dichotomy of control became the opening lesson of his teaching. He divided everything in life into two categories: things "up to us" (eph' hēmin) and things "not up to us" (ouk eph' hēmin).

Up to us: Our judgments, intentions, desires, aversions—in short, our own mental operations and voluntary actions. These are completely within our control.

Not up to us: Our bodies, possessions, reputations, positions, and external outcomes. These depend partly on our efforts but also on countless factors beyond our influence.

Epictetus was uncompromising about this distinction. He argued that most human misery comes from confusing the two categories—from trying to control what we can't control, and from neglecting what we can.

When you need a specific outcome to feel okay, you've handed your wellbeing to forces outside your control. When you require other people to behave a certain way, you've made your peace contingent on their choices. When you demand certainty before you can act, you've made yourself dependent on an impossibility.

The Stoic solution isn't passivity. It's redirected energy. Pour your effort into what you can actually influence—your preparation, your choices, your responses—and release attachment to what you can't.

• • •

The Archer Metaphor

The Roman philosopher Cicero captured this principle with an image that has endured for two millennia.

Imagine an archer competing in a tournament. What is the archer's job? To aim well. To draw the bow with proper form. To release at the right moment. To do everything within their power to send the arrow toward the target.

But once the arrow leaves the bow, it's no longer the archer's. A gust of wind might blow it off course. The target might shift. Another arrow might knock it aside. The archer cannot control the flight of the arrow after release—only the quality of the aim before it.

If the archer's goal is "hit the target," they've set themselves up for anxiety. They've attached their success to something partly outside their control. But if the archer's goal is "aim as well as I possibly can," they can succeed regardless of where the arrow lands.

They can walk away knowing they did their job, even if the outcome wasn't what they hoped.

This isn't semantic game-playing. It's a fundamental reorientation of what success means.

The archer who aims well and misses has succeeded at the only thing that was ever theirs to succeed at. The archer who aims poorly and hits—through luck or favorable conditions—has actually failed at the task, despite the apparent victory.

Apply this to your presentation. Your job was to prepare thoroughly, deliver effectively, and represent your ideas with clarity. If you did those things, you succeeded—regardless of the committee's decision. If you didn't do those things but got approved anyway, you got lucky. You didn't succeed at anything you controlled.

This reframe doesn't eliminate preference. The archer still wants to hit the target. You still want the proposal approved. But wanting an outcome is different from needing it. And that difference changes everything about how you experience the waiting period.

• • •

Modern Psychology Catches Up

For centuries, the dichotomy of control remained philosophical wisdom. Then modern psychology started measuring it.

Acceptance and Commitment Therapy (ACT), developed in the 1980s, centers on a concept called psychological flexibility—the ability to stay present, open to experience, and engaged in values-driven action regardless of internal discomfort or external circumstances.

The ACT model identifies six core processes that build psychological flexibility. Several map directly onto Stoic principles:

Acceptance: Embracing thoughts and emotions as they are, rather than fighting to eliminate them. This parallels the Stoic practice of accepting what's not in our control.

Cognitive defusion: Creating distance from thoughts, seeing them as mental events rather than literal truths. This parallels the Stoic practice of examining impressions before assenting to them.

Values-driven action: Acting according to what matters most to you, regardless of discomfort. This parallels the Stoic focus on virtue and character over outcomes.

Committed action: Taking concrete steps aligned with values, even when outcomes are uncertain. This parallels the Stoic emphasis on doing your part while releasing attachment to results.

Research on psychological flexibility shows significant correlations with wellbeing. A 2022 study during Stoic Week found that participants who reported "controlling the controllables" showed a .51 correlation with flourishing. Those who reported "letting go of things I cannot control" showed a .47 correlation with positive emotional balance.

The ancient framework and the modern data converge: distinguishing what you control from what you don't is measurably associated with better psychological outcomes.

• • •

The Reserve Clause

The Stoics weren't naive optimists. They didn't pretend everything would work out. They used a mental strategy called the "reserve clause"—adding a qualifier to every intention.

When Marcus Aurelius set a goal, he would add "fate permitting" or "if nothing prevents it." This wasn't pessimism. It was realism. He was acknowledging that his intention was within his control, but the outcome depended on factors beyond it.

"I will finish this project—fate permitting."

"I will have a productive conversation with my colleague—if nothing prevents it."

"I will secure this client—God willing."

The reserve clause does two things. First, it prepares you psychologically for obstacles. You've already acknowledged that things might not go as planned, so when they don't, you're not blindsided. Second, it clarifies what's actually yours: the intention, the effort, the attempt. Not the result.

This might sound like hedging or lowering expectations. It's not. Marcus Aurelius ran an empire. He didn't achieve that by being passive or unambitious. The reserve clause doesn't reduce effort—it focuses effort on what effort can actually accomplish.

You can work as hard as possible to close the deal while acknowledging that the client's decision isn't yours to make. You can prepare obsessively for the interview while accepting that the hiring committee has their own criteria. You can give everything to the relationship while recognizing that your partner is a separate person making their own choices.

Ambition without attachment. That's the Stoic way.

• • •

Viktor Frankl and the Last Freedom

In 1942, Viktor Frankl, a Jewish psychiatrist, was deported to Nazi concentration camps. Over the next three years, he lost his parents, his brother, and his pregnant wife. He was stripped of everything—possessions, profession, physical freedom, human dignity.

In that abyss, Frankl discovered something that changed his understanding of human psychology. He wrote about it later in *Man's Search for Meaning*:

> *Everything can be taken from a man but one thing: the last of the human freedoms—to choose one's attitude in any given set of circumstances, to choose one's own way.*

Frankl watched prisoners in identical circumstances respond differently. Some gave up and died. Others found meaning, helped fellow prisoners, and survived against statistical odds. The external situation was the same. The internal response varied completely.

This wasn't positive thinking. Frankl didn't pretend the camps were good or that suffering was an illusion. He faced the horror directly. But he recognized that even in the worst circumstances imaginable, one choice remained: how to interpret and respond to those circumstances.

The dichotomy of control, tested in the most extreme conditions possible.

Frankl survived. He went on to develop logotherapy, a psychiatric approach centered on meaning-making. His work influenced millions. And it began with the same insight Epictetus taught nineteen centuries earlier: your response is yours, even when nothing else is.

• • •

The Three Control Shifts

With this foundation in place, here are three shifts that apply the dichotomy of control to situations you face constantly.

SHIFT 4

FROM: *"I need this to work out."*

TO: *"I need to work on this; the outcome is not mine."*

This is the outcome attachment shift. It targets the belief that your wellbeing depends on a specific result.

When you think "I need this job," you're not just expressing preference. You're making your emotional state contingent on a decision someone else will make. If you get the job, relief. If you don't, devastation. Either way, you've surrendered your equilibrium to factors beyond your control.

The shift relocates your focus. "I need to work on this" keeps the energy directed at what you can actually do—preparation, performance, follow-up. "The outcome is not mine" releases the grip on what you can't.

This doesn't mean you stop caring about outcomes. You can strongly prefer a result while accepting that you don't control it. The archer wants to hit the target. But the archer's job is aiming, not hitting. Your job is working, not guaranteeing results.

Notice the practical difference. When you need it to work out, you're anxious during the process and devastated if it doesn't. When you need to work on it, you're focused during the process and at peace afterward—because you did your job regardless of how it turned out.

SHIFT 5

FROM: *"They should behave differently."*

TO: *"I can only control my response."*

This is the other-people shift. It targets the belief that other people's behavior should conform to your expectations.

"They should" is a setup for frustration. Your colleague should meet deadlines. Your partner should understand you. Your boss should recognize your contributions. These statements feel reasonable. They might even be accurate assessments of professional or relational norms.

But "should" doesn't change behavior. Other people operate on their own motivations, their own histories, their own interpretations of the world. You can influence them—sometimes. You can request, persuade, incentivize, set boundaries. What you cannot do is control their choices.

When you think "they should behave differently," you're demanding control over something not up to you. The inevitable result is frustration, because other people will keep doing what they do regardless of your demands.

The shift redirects to what's actually yours. They did what they did. Now what will you do? How will you respond? What's your next move? These questions put the focus back on your agency.

Marcus Aurelius, who dealt with difficult people constantly as emperor, wrote reminders to himself: "Begin each day by telling yourself: Today I shall be meeting with interference, ingratitude, insolence, disloyalty, ill-will, and selfishness." He wasn't being cynical. He was being realistic—and freeing himself from the expectation that people would behave otherwise.

When you expect difficult behavior, you're not thrown by it. You can plan your response in advance. You stay in control of the only thing you ever controlled: yourself.

SHIFT 6

FROM: *"I must eliminate uncertainty."*

TO: *"I can act effectively within uncertainty."*

This is the certainty shift. It targets the belief that you need to know how things will turn out before you can move forward.

Uncertainty is uncomfortable. Your brain treats it as a threat. It generates anxiety precisely because it can't predict what's coming. The natural response is to seek more information, more assurance, more control over the unknown.

But certainty is largely an illusion. You don't know how the market will move, how the conversation will go, how your health will hold up, how the relationship will develop. You can make educated guesses. You can't know.

Demanding certainty before action creates paralysis. You wait for information that never arrives. You delay decisions hoping for clarity that doesn't come. You stay stuck because moving forward feels too risky without guarantees you'll never get.

The shift accepts uncertainty as a permanent condition. You will never have complete information. You will always be acting on incomplete data. The question isn't whether to act despite uncertainty—that's the only option—but whether to act skillfully within it.

"I can act effectively within uncertainty" doesn't mean recklessness. It means making the best decision you can with the information available, accepting that new information might change things, and moving forward anyway. It means becoming comfortable with not knowing while still doing.

The Stoics called this living with a "reserve clause." You act with full commitment while acknowledging that outcomes aren't guaranteed. You do your part without demanding that the universe guarantee the result.

• • •

Putting It Together

The dichotomy of control gives you a sorting mechanism for everything you encounter.

When you feel anxiety, frustration, or desperation, ask: Am I trying to control something that isn't up to me? Am I demanding an outcome? Am I requiring other people to change? Am I insisting on certainty?

If the answer is yes, you've located the source of your suffering. Not the external situation—your orientation toward it. You're grasping at something that was never yours to hold.

The solution isn't giving up. It's redirecting. What *is* up to you in this situation? Your preparation. Your effort. Your response. Your character. Your choice of what to do next.

Focus there. Release the rest. Not because the rest doesn't matter—it might matter enormously—but because holding on to it gives you nothing except anxiety. The outcome doesn't change based on how much you worry about it. Your wellbeing does.

• • •

TAKE ACTION: YOUR NEXT STEPS

Reflection

Think of a situation that's currently causing you stress or anxiety. Work through these questions:

1. What outcome are you attached to? What specific result do you feel you "need" to happen? Name it clearly.

2. Is this outcome within your control? Be honest. Can you guarantee this result through your efforts alone, or does it depend on other people, circumstances, or factors you can't determine?

3. What is within your control? List the specific actions, preparations, and responses that are entirely up to you. These are your domain.

4. What would change if you released the outcome? If you fully committed to what you control and accepted that the outcome isn't yours, how would your experience of this situation shift?

Implementation

Practice the reserve clause. For the next week, when you set an intention or make a plan, add "fate permitting" or "if nothing prevents it" mentally. Notice how this changes your relationship to the goal.

Identify your "shoulds." Pay attention to when you think "they should" about other people. Write down three instances this week. For each one, ask: What can I control here? What's my response going to be?

Take one action despite uncertainty. Identify something you've been postponing because you don't have enough information or certainty. Take one concrete step forward this week, accepting that you're acting on incomplete data. Notice what happens.

Troubleshooting

"This sounds like giving up. I don't want to stop caring about outcomes."

The dichotomy of control doesn't eliminate preference. You can strongly want something while accepting that you don't control whether you get it. The archer wants to hit the target. Wanting isn't the problem—needing is. You release attachment, not aspiration.

"But sometimes I can influence outcomes. Isn't that a form of control?"

Influence is real. You can increase probabilities. But influence isn't control. You can do everything right and still not get the result, because other factors are at play. Focus on maximizing your influence—that's within your control. Accept that influence doesn't guarantee outcomes.

"What if the thing I can't control is genuinely important—like my health or my child's safety?"

The stakes don't change the principle. Important things that aren't in your control are still not in your control. You can take every reasonable precaution for your health and still get sick. You can be the best parent possible and your child can still face challenges you can't prevent. The dichotomy doesn't make this easy to accept. But fighting against reality doesn't change it—it only adds suffering to difficulty.

• • •

You now have two foundational frameworks: the Event-Interpretation-Response chain from Chapter 1, and the Dichotomy of Control from this chapter. Together, they form the operating system for everything that follows.

You know that your experience is shaped by interpretation, not just events. You know that much of what you try to control isn't up to you. These insights free you to focus where focus matters: on your own judgments, choices, and responses.

With this foundation in place, we're ready to apply these principles to specific domains. Part Two begins with career—where cognitive distortions about failure, criticism, uncertainty, status, and meaning show up constantly, and where the cost of distorted thinking is measured in missed opportunities, unnecessary suffering, and stalled progress.

Let's get specific.

PART TWO

CAREER SHIFTS

CHAPTER 3: Reframing Failure and Setback

The cognitive shifts that separate people who recover quickly from those who spiral

> *I have not failed. I've just found 10,000 ways that won't work.*
>
> — Thomas Edison

You didn't get the promotion. The project fell apart. The client said no. The launch flopped. The deal collapsed at the last minute.

Now what?

Two paths diverge from this moment. One leads to a spiral: replaying what went wrong, questioning your competence, catastrophizing about your career, withdrawing from risk, performing worse because you're now afraid to fail again. The other leads to recovery: extracting the lesson, adjusting the approach, moving forward with the same ambition but better information.

The difference between these paths isn't resilience as a personality trait. It's not grit or mental toughness or any other fixed quality you either have or don't. The difference is interpretation. The people who recover quickly interpret failure differently than the people who spiral.

This chapter is about learning that interpretation.

• • •

The Overgeneralization Trap

When something goes wrong, your brain doesn't contain the damage. It spreads it.

A single failed project becomes evidence that you're bad at your job. One rejected proposal proves you'll never succeed in this field. A mistake in a presentation means you can't handle pressure.

This is overgeneralization—one of the most common and most destructive cognitive distortions. It takes a single data point and extrapolates it into a universal truth. The pattern shows up linguistically. Listen for the words "always," "never," "everything," and "nothing."

"I always mess up under pressure."

"I'll never be good at this."

"Everything I try fails."

"Nothing works out for me."

These statements feel true in the moment. They're not. They're vast generalizations built on limited evidence, usually colored by the emotional intensity of the recent failure.

Consider the logic. You gave one presentation that didn't land well. From this single event, you conclude that you're constitutionally incapable of presenting effectively. But what about the presentations that went fine? What about the ones that went well? Overgeneralization discounts all contrary evidence. It cherry-picks the failure and treats it as definitive.

The reality is that a single event tells you almost nothing about your underlying capabilities. Performance varies. Circumstances vary. You might have been underprepared, or facing an unusually tough audience, or dealing with factors outside your control. One outcome doesn't establish a pattern. It takes many data points to identify an actual trend.

When you catch yourself generalizing from a single failure, ask: Would I convict someone in court based on this much evidence? The answer is no. You'd demand more data. Apply the same standard to yourself.

• • •

Why Single Events Don't Define Capabilities

Here's a thought experiment. Imagine a basketball player who misses a free throw. Does that missed shot prove they're a bad free-throw shooter?

Obviously not. Free-throw percentage is calculated over hundreds of attempts. A single miss is noise—random variation within a larger pattern. Even the best shooters miss sometimes. Even the worst shooters make some.

Now imagine that same player misses a free throw in a crucial playoff game. Does that prove they "choke under pressure"?

Still no. You'd need data from many high-pressure situations to draw that conclusion. One missed shot—even a consequential one—is still just one data point.

Your professional capabilities work the same way. Your ability to lead a team, close a deal, write clearly, manage a project, or handle criticism is demonstrated across dozens or hundreds of instances, not in any single moment.

When you fail at something once, you've learned that you can fail at it. You haven't learned that you're incapable of succeeding. The difference is enormous.

Professional sports provide a useful lens here because the statistics are public and comprehensive. Baseball players who hit .300—considered excellent—fail to get a hit 70% of the time. Venture capitalists expect most of their investments to fail. Salespeople celebrate closing rates of 20-30%. In all these domains, failure is the norm, not the exception.

But when you're inside your own career, a single failure can feel catastrophic. That's because you're not tracking your overall batting average. You're fixating on the at-bat you just struck out on.

Zoom out. Your capabilities are defined by the aggregate, not the instance.

• • •

Fixed vs. Growth Mindset—Through a Cognitive Lens

Psychologist Carol Dweck's research on mindsets has become famous, sometimes to the point of oversimplification. But the core insight maps directly onto cognitive distortions.

A fixed mindset interprets failure as evidence of permanent limitation. "I failed, therefore I lack the ability." Ability is seen as static—you either have it or you don't. Failure reveals that you don't.

A growth mindset interprets failure as evidence of an incomplete process. "I failed, therefore my approach needs adjustment." Ability is seen as developable—you build it through effort, feedback, and iteration. Failure provides information for the next attempt.

From a cognitive-behavioral perspective, the fixed mindset is a cluster of distortions: overgeneralization (one failure means permanent incapacity), personalization (the failure reflects my essential nature), and all-or-nothing thinking (either I have the ability or I don't).

The growth mindset is actually the more accurate interpretation. Abilities do develop. Skills do improve with practice. Strategies can be refined. The fixed mindset isn't just unhelpful—it's wrong about how competence works.

But here's what Dweck's popular framing sometimes misses: you can't just decide to have a growth mindset. If you're in the habit of interpreting failures through a fixed lens, that habit is automatic. You'll think "I'm not good enough" before you think anything else.

The solution is the same as with other cognitive distortions: catch the interpretation, examine it, and apply a more accurate reframe. The shifts in this chapter give you specific reframes to practice. With repetition, the growth-oriented interpretation becomes more automatic. You're not just adopting a mindset—you're rewiring a thinking pattern.

• • •

The Evidence-Gathering Technique

When you fail, your brain generates an interpretation instantly. "I'm not cut out for this." "I always choke." "This proves what I suspected about myself."

These interpretations feel like conclusions, but they're actually hypotheses. And hypotheses can be tested.

The evidence-gathering technique comes from cognitive behavioral therapy. It works like this:

Step 1: State the belief. Write down the interpretation your mind generated. Be specific. "I failed this presentation because I'm bad at public speaking."

Step 2: List evidence that supports the belief. What data points support this interpretation? "I stumbled over my words. I forgot a key point. The audience seemed disengaged."

Step 3: List evidence that contradicts the belief. What data points suggest the interpretation might be wrong or incomplete? "I've given successful presentations before. I was underprepared this time due to a deadline crunch. The audience was also distracted by a fire drill earlier."

Step 4: Generate an alternative interpretation. Based on all the evidence, what's a more balanced conclusion? "This presentation didn't go well because I was underprepared. When I have adequate preparation time, I present effectively."

The technique works because it forces you to consider evidence you'd otherwise ignore. In the grip of a failure, your mind filters for information that confirms your worst interpretation. Evidence-gathering counters that filter by requiring you to actively seek contrary data.

Most of the time, you'll find that the catastrophic interpretation doesn't hold up. The evidence for "I'm bad at this" is usually much weaker than it feels. And the evidence for "this attempt had specific, addressable problems" is usually much stronger.

• • •

The Four Failure Shifts

With the framework in place, here are four specific shifts for reframing failure and setback.

SHIFT 7

FROM: *"I failed."*

TO: *"This attempt produced data."*

The word "failure" carries weight. It implies finality. A verdict. Case closed.

But most failures aren't final. They're iterative. They're points in a process, not conclusions. The attempt didn't produce the result you wanted—it produced information about what works and what doesn't.

Thomas Edison's famous quote about finding 10,000 ways that don't work isn't just motivational fluff. It's an accurate description of experimental method. Every attempt that doesn't succeed tells you something. It narrows the solution space. It reveals constraints you didn't know existed. It generates data.

When you reframe "I failed" as "this attempt produced data," you shift from judgment to analysis. The question changes from "What does this say about me?" to "What does this tell me about the problem?"

The data might reveal that your approach was flawed. Good—now you can adjust it. The data might reveal that the timing was wrong. Good—now you can recalibrate. The data might reveal that this particular goal isn't achievable with current resources. Good—now you can make an informed decision about whether to persist, pivot, or quit.

All of that is valuable. None of it is available if you stop at "I failed" and stay there.

SHIFT 8

FROM: *"This shouldn't have happened."*

TO: *"This happened; now what?*

"This shouldn't have happened" is a protest against reality. It's understandable—when something goes wrong, we want to reject it. But the protest changes nothing. The failure has already occurred. Insisting it shouldn't have is just adding suffering to difficulty.

The Stoics were uncompromising about this. Marcus Aurelius wrote: "Accept the things to which fate binds you, and love the people with whom fate brings you together." Acceptance wasn't passive resignation—it was clearing the way for effective action.

When you're stuck in "this shouldn't have happened," your energy goes toward resistance, resentment, and rumination. You replay the events. You catalog the injustices. You argue with the past.

When you shift to "this happened; now what?" your energy redirects toward the future. The situation is what it is. The question is what you're going to do about it.

This shift doesn't require you to like what happened or pretend it wasn't bad. You can acknowledge that the outcome was painful, disappointing, or unfair. But acknowledgment is different from resistance. You can accept that something happened while working to make sure it doesn't happen again.

"Now what?" is the only productive question after failure. Everything else is delay.

SHIFT 9

FROM: *"This proves I'm not good enough."*

TO: *"This proves I tried something difficult."*

This shift directly attacks the personalization distortion—the tendency to interpret failures as statements about your essential worth or capability.

"This proves I'm not good enough" treats failure as evidence against you. It assumes that competent people don't fail, that failure reveals inadequacy, that the struggle itself is shameful.

But failure is most common when you're attempting something difficult. Easy tasks don't produce much failure because they don't require much from you. The failures that hurt—the ones that make you question yourself—usually come from stretching beyond your current capacity.

That stretch is not evidence of inadequacy. It's evidence of ambition.

The person who never fails is the person who never attempts anything challenging. They're staying safely within their existing competence, avoiding any risk of coming up short. They look successful because they've rigged the game—only playing matches they know they can win.

Is that the model you want to follow?

When you fail at something difficult, you've demonstrated willingness to take on challenges that might beat you. You've revealed something about your courage, not your incompetence. The failure proves you were in the arena.

"This proves I tried something difficult" reframes failure as a badge of effort rather than a mark of shame. The interpretation is just as accurate as the negative one—and far more useful for what comes next.

SHIFT 10

FROM: *"I wasted all that time."*

TO: *"I learned what doesn't work, which is progress."*

After a failed project or initiative, the sunk cost feels enormous. All those hours. All that effort. All for nothing.

But was it for nothing?

Learning what doesn't work is genuinely valuable—sometimes more valuable than success. When something works, you don't always know why it worked. When something fails, you have to figure out what went wrong. That diagnostic process produces insight.

Consider the alternative. If you hadn't tried, you'd still be wondering whether the approach would work. Now you know. That knowledge isn't nothing. It's progress.

The mental model here is elimination. Every approach you try and discard eliminates a path you might otherwise have wasted future time on. You're narrowing toward what works by systematically identifying what doesn't.

Scientists understand this intuitively. A failed experiment isn't wasted time—it's data that shapes the next experiment. The null result (this approach doesn't produce the expected effect) is still a result. It tells you something you didn't know before.

"I learned what doesn't work" reframes the investment as productive rather than wasted. The time is gone either way. What you do with the learning determines whether it was well spent.

• • •

Putting It Together

When failure hits, you have a sequence to run.

First, notice the automatic interpretation. What story is your mind telling about what this failure means? "I failed" / "This shouldn't have happened" / "I'm not good enough" / "I wasted my time."

Second, apply the relevant shift. Reframe to the more accurate interpretation. "This attempt produced data" / "This happened; now what?" / "This proves I tried something difficult" / "I learned what doesn't work."

Third, examine the evidence. Does the catastrophic interpretation actually hold up? What evidence contradicts it? What would a more balanced assessment look like?

Fourth, extract the learning. What specifically can you take from this failure? What will you do differently next time? What did you discover that you didn't know before?

Fifth, move forward. The failure is in the past. The learning is with you. Now apply it.

This process gets faster with practice. Eventually, the reframes become more automatic. You'll still feel the sting of failure—that's human. But you'll spend less time spiraling and more time recovering.

• • •

TAKE ACTION: YOUR NEXT STEPS

Reflection

Think of a recent professional failure or setback—something that still bothers you. Work through these questions:

1. What interpretation did your mind generate? Write down the exact thought. "I failed because..." / "This proves..." / "I should have..."

2. What cognitive distortion is operating? Is this overgeneralization? Personalization? All-or-nothing thinking? Name the pattern.

3. What evidence contradicts your interpretation? Force yourself to list at least three data points that suggest a different conclusion.

4. What's a more balanced interpretation? Rewrite the story of this failure in a way that accounts for all the evidence.

5. What did you learn? What specific insight came from this failure that you can apply going forward?

Implementation

Track your failure language this week. When something goes wrong, notice what words you use. Are you saying "always," "never," "everything," "nothing"? These signal overgeneralization. Practice replacing them with specific, bounded language. Not "I always mess up presentations" but "This presentation didn't land."

Practice the data reframe in real time. The next time something doesn't work out, before you do anything else, say to yourself: "This attempt produced data. What's the data?" Force yourself to extract information before you spiral into self-criticism.

Keep a "lessons learned" log. For one month, record every professional failure or setback along with one specific lesson you extracted. At the end of the month, review the log. You'll likely find that failures you felt terrible about produced valuable insights.

Troubleshooting

"The failure really was my fault. Shouldn't I take responsibility?"

Absolutely. Taking responsibility and spiraling are different things. You can acknowledge that you made mistakes without concluding that you're fundamentally incapable. "I was underprepared for that meeting" is taking responsibility. "I'm terrible at my job" is overgeneralization. Own your part. Then move forward.

"But this failure has real consequences. My reputation is damaged. I might get fired."

The consequences are real. The reframes don't make them disappear. What the reframes do is help you respond effectively rather than spiraling. If your reputation is damaged, what's the best way to rebuild it—by catastrophizing or by demonstrating competence on the next opportunity? If you might get fired, what's the best use of your energy—ruminating on what went wrong or positioning for what comes next?

"Some failures really do reveal that you're not suited for something. How do you know the difference?"

Fair point. Not every failure is a learning opportunity; some are genuine signals that a path isn't right for you. The difference is pattern recognition. A single failure tells you very little. Repeated failures in the same area, despite genuine effort and adjustment, start to tell you something. Even then, the question isn't "Am I bad at this?" but "Is continued investment in this area the best use of my time and energy?" That's a strategic question, not an identity question.

• • •

Failure is inevitable in any ambitious career. The question is never whether you'll fail but how you'll interpret failure when it comes.

The shifts in this chapter give you a different relationship with setback. Not denial—acknowledgment. Not shame—analysis. Not paralysis—forward motion.

In the next chapter, we'll tackle a related but distinct challenge: how to handle criticism and negative feedback. Failure is what happens when things don't work out. Criticism is when other people tell you things didn't work out—often in ways that feel personal, unfair, or threatening. The cognitive skills are related, but the situations require their own specific reframes.

CHAPTER 4: Reframing Criticism and Feedback

How to process negative input without defensive spiraling or self-attack

> *If anyone tells you that a certain person speaks ill of you, do not make excuses about what is said of you but answer, "He was ignorant of my other faults, else he would not have mentioned these alone."*
>
> — Epictetus, Enchiridion

Your manager sends an email: "Can we talk about the Henderson report? I have some concerns."

What happens in your body? For most people, something tightens. The stomach. The chest. The jaw. Before you've even read what the concerns are, before you know whether they're minor or major, valid or mistaken—your threat-detection system has activated.

Now your mind starts generating interpretations. She thinks I did a bad job. She's disappointed in me. Maybe she's building a case to let me go. I knew I shouldn't have rushed it. Everyone probably thinks I'm incompetent.

By the time you walk into the meeting, you're already defensive. You're already rehearsing justifications. You're already filtering whatever she says through the lens of "I'm being attacked." And that lens will shape everything—how you hear her words, how you respond, and what happens to the working relationship afterward.

This chapter is about a different way to receive criticism. Not a way that makes it pleasant—criticism rarely is. But a way that lets you extract value from negative feedback without spiraling into self-attack or defensive warfare.

• • •

The Personalization Distortion

When someone critiques your work, your brain has a tendency to translate that critique into a statement about you.

They said the report was disorganized. Your brain hears: They think I'm disorganized.

They said the presentation lacked focus. Your brain hears: They think I lack focus.

They said the code had bugs. Your brain hears: They think I'm a bad programmer.

This is the personalization distortion—interpreting feedback about outputs as judgments about your character, competence, or worth. The gap between "the work has a problem" and "I have a problem" collapses. They become the same thing.

But they're not the same thing.

A report can be disorganized without you being a disorganized person. You might have been rushed, unclear on requirements, or working outside your expertise. A presentation can lack focus because the brief was vague, the stakeholders wanted contradictory things, or you tried to cover too much. The output reflects many variables, of which your underlying capability is only one.

When you personalize feedback, you make the conversation about your identity instead of the work. That raises the emotional stakes dramatically. Suddenly you're not discussing how to improve a document—you're defending your fundamental competence. No wonder the meetings feel threatening.

The first step in handling criticism better is catching the personalization as it happens. When you notice yourself translating feedback about work into feedback about you, pause. Name the distortion. Remind yourself: they're talking about an output, not making a declaration about my worth.

• • •

Mind-Reading Errors

Personalization has a partner: mind-reading. While personalization turns feedback about work into feedback about you, mind-reading fills in what the other person "really" thinks—usually with worst-case interpretations.

Your colleague says: "This section could be clearer."

You hear: She thinks I'm a poor communicator.

What she actually thinks: This section could be clearer.

You have no idea whether she's concluded anything about your general communication abilities. She might think this section could be clearer while also thinking you're one of the best writers on the team. She might think this section could be clearer because she's having a tired day and needs simpler prose. She might not be thinking about you at all—just about the document.

Mind-reading assumes that you have access to other people's internal evaluations. You don't. You have access to what they said. Everything else is your projection.

This matters because your emotional response follows from what you think they think, not from what they actually said. If you're responding to "she thinks I'm incompetent" rather than "this section could be clearer," you're responding to a fantasy. Your defensiveness, your hurt, your anxiety—all directed at something that exists only in your interpretation.

The fix is deceptively simple: respond to what was actually said. When someone gives you feedback, take the words at face value. Resist the urge to decode hidden meanings or infer broader judgments. They said the section could be clearer. That's the data. Everything else is story.

• • •

Cognitive Defusion: Separating You from the Feedback

Acceptance and Commitment Therapy offers a concept that's directly applicable here: cognitive defusion.

Fusion is when you're merged with a thought—when the thought feels like reality rather than a mental event. If someone criticizes your work and you think "I'm incompetent," fusion means you experience that thought as truth. You don't observe the thought; you become the thought. Your incompetence feels like an established fact.

Defusion creates separation between you and the thought. You observe the thought as a thought—a passing mental event, not an objective reality. You might think "I'm having the thought that I'm incompetent." The content is the same, but the relationship to the content has changed. You're watching the thought rather than being consumed by it.

This applies directly to receiving criticism. When feedback triggers negative self-evaluation, you can defuse from that evaluation.

Fused: "I'm a failure."

Defused: "I'm noticing a thought that I'm a failure."

Fused: "They think I'm incompetent."

Defused: "My mind is telling me that they think I'm incompetent."

The defused statements might sound awkward. That's intentional. The linguistic distance creates psychological distance. It reminds you that thoughts are not facts—they're events happening in your mind, and you can observe them without being controlled by them.

In practice, defusion doesn't make criticism painless. The thoughts still arise. But you have a different relationship to them. They're not pronouncements from on high. They're mental weather—something to notice, maybe learn from, but not something that defines reality.

• • •

The Stoic Practice of Objective Representation

The Stoics had their own version of defusion. They called it "objective representation" or describing things as they really are, stripped of value judgments and emotional coloring.

Marcus Aurelius practiced this constantly. When faced with something that seemed threatening or insulting, he would describe it in bare, physical terms. Wine is fermented grape juice. A royal robe is sheep's wool dyed with shellfish blood. Sexual intercourse is friction and spasm.

The practice sounds almost comically reductive. But that's the point. By stripping away the emotional significance we layer onto things, we see them more clearly. And when we see them clearly, we respond more rationally.

Apply this to criticism. When someone criticizes your work, what actually happened?

Objective representation: A person made sounds with their mouth (or typed characters on a screen) that described perceived shortcomings in a document I produced.

That's it. That's the event. Everything else—the threat to your identity, the implication about your competence, the prediction about your career—is added by your interpretation.

This isn't about dismissing criticism or pretending it doesn't matter. It's about seeing the criticism accurately so you can respond to it usefully. The objective event is that someone identified issues with your work. That's information. What you do with the information is up to you.

When you mentally strip criticism down to its bare form, it becomes less overwhelming. It's not an assault on your personhood. It's data about an output. Data you can use.

• • •

The Source of Value in Criticism

Here's an uncomfortable truth: criticism is often where the most valuable information lives.

Praise tells you what you're already doing well. That's nice, but it doesn't help you improve. Criticism points to gaps—places where your work falls short, blind spots you can't see yourself, weaknesses that limit your effectiveness.

If your goal is growth, criticism is more useful than praise.

This doesn't mean all criticism is valid or well-delivered. Some critics are wrong. Some are motivated by their own insecurities. Some deliver feedback in ways that are needlessly harsh or personal. You don't have to accept every criticism as true or fair.

But even poorly delivered criticism sometimes contains signal. Even unfair criticism sometimes points to something real. The question is whether you can extract the useful information without being derailed by the delivery.

The shifts that follow are designed to help you do exactly that—to receive criticism in a way that lets you access its value while protecting yourself from unnecessary psychological damage.

• • •

The Four Criticism Shifts

SHIFT 11

FROM: *"They think I'm incompetent."*

TO: *"They think this work could be better."*

This shift attacks mind-reading directly.

When someone critiques your work, you have no idea what they think about you as a person or professional. You're inferring that from limited data—and your inference is almost certainly skewed negative.

What you do know is what they said. They identified issues with the work. That's a statement about the work, not about your competence.

"They think this work could be better" stays grounded in the observable. It's a reasonable interpretation of what criticism means—someone sees room for improvement in an output. That's it. No hidden message about your capability. No implied judgment about your career prospects. Just: this work could be better.

Can you receive that information? Can you hear "this work could be better" without translating it into "I'm not good enough"?

The shift matters because the two interpretations produce completely different responses. If they think you're incompetent, you need to defend yourself. If they think the work could be better, you need to improve the work. One leads to conflict. The other leads to progress.

SHIFT 12

> **FROM:** *"I need everyone to approve."*
>
> **TO:** *"I need my work to improve."*

This shift addresses the approval trap.

Some people are so sensitive to criticism because they're playing an impossible game: trying to get everyone to approve of their work all the time. Every piece of negative feedback is a failure of that mission. Every critic is evidence that they haven't achieved universal acceptance.

But universal approval isn't achievable. People have different standards, different preferences, different contexts. Work that one person loves, another finds mediocre. Work that impresses this stakeholder disappoints that one. You cannot produce anything of significance without someone finding fault.

If your goal is approval, you're at the mercy of every critic. One negative voice can ruin your day, because it represents failure of your objective.

If your goal is improvement, critics become resources. They're pointing to areas where your work can get better. You don't need them to approve—you need them to give you information. And criticism is information.

The shift relocates your focus from external validation to internal progress. Instead of asking "Do they like it?" you ask "What can I learn from this?" The first question makes you dependent on others' opinions. The second keeps the power in your hands.

SHIFT 13

> **FROM:** *"This feedback defines me."*
>
> **TO:** *"This feedback describes one output."*

This shift attacks the scope problem.

Criticism of one piece of work feels like it defines you entirely. Your identity, competence, and value all seem to hang on this single judgment. But that's a category error. The feedback is about an output—a report, a presentation, a project, a decision. It's not about you.

You are not your work. You produced the work. The work exists separate from you. Criticism of the work is criticism of something you made, not something you are.

This distinction matters practically. If the feedback defines you, then improving feels impossible—you'd have to change your fundamental nature. If the feedback describes one output, then improving is straightforward—you make the next output better.

Consider how you'd think about criticism of a meal you cooked. "This soup is too salty" doesn't mean you're a bad person or that you should never cook again. It means this particular soup has too much salt. You can fix that next time. The feedback is bounded to the output.

Your professional work deserves the same bounded interpretation. This report has organizational issues. This presentation lacked focus. This code had bugs. Specific outputs with specific problems. Not declarations about your essential nature.

SHIFT 13

FROM: *"They're attacking me."*

TO: *"They're addressing the work."*

This shift reframes the intent behind criticism.

When criticism feels like an attack, your body responds accordingly. Fight-or-flight activates. You become defensive, aggressive, or withdrawn. The conversation becomes adversarial. Whatever value the feedback might have contained gets lost in the conflict.

But most criticism isn't an attack. Most critics aren't trying to hurt you—they're trying to improve the work. They have a stake in the outcome. They want the project to succeed, the document to be clear, the product to work well. Their criticism comes from engagement, not hostility.

Even when criticism is delivered poorly—with frustration, impatience, or condescension—the underlying motive is usually "make this better," not "destroy this person."

"They're addressing the work" assumes good intent, or at least neutral intent. It frames the critic as someone focused on the output, not someone targeting you personally. This framing might not always be accurate—some people really are hostile. But it's accurate often enough that using it as your default interpretation will serve you well.

When you assume you're being attacked, you attack back. When you assume the work is being addressed, you address the work. The second response is almost always more productive.

• • •

Putting It Together

When criticism arrives, you have a sequence available.

Notice the physical response. Feel the tightening, the defensiveness, the threat activation. Recognize that your body is responding to perceived danger.

Catch the interpretation. What story is your mind generating? "They think I'm incompetent." "Everyone will see I'm a fraud." "This proves I'm not good enough." Name the interpretation.

Apply defusion. "I'm having the thought that they think I'm incompetent." Create distance between you and the mental event.

Use objective representation. What actually happened? Someone identified issues with an output. That's the event. Everything else is addition.

Apply the relevant shift. Move from the distorted interpretation to the more accurate one. They're not attacking you—they're addressing the work. The feedback describes one output, not your entire identity.

Extract the value. What useful information does the criticism contain? What can you do with it? What might you do differently?

This sequence won't make criticism enjoyable. It will make criticism manageable. You'll spend less time defending, ruminating, and spiraling. You'll spend more time learning and improving.

• • •

TAKE ACTION: YOUR NEXT STEPS

Reflection

Recall a recent piece of criticism that stung—feedback that felt personal, unfair, or threatening. Work through these questions:

1. What was the actual feedback? Write down only what was said, as close to the exact words as you can remember. No interpretation, no inferred meaning. Just the words.

2. What interpretation did you add? What conclusions did you draw about what they really meant, what they thought of you, or what the feedback implied about your future?

3. What mind-reading did you do? What did you assume about their internal evaluation of you that went beyond what they actually said?

4. How would you describe the feedback using objective representation? Strip it down to the bare, factual description. What happened, physically?

5. What useful information was contained in the feedback? Setting aside how it was delivered or how it felt, was there anything accurate in the criticism that you could act on?

Implementation

Practice the defusion language. This week, when you receive any criticism—even minor—practice inserting "I'm having the thought that..." before your automatic interpretation. "I'm having the thought that she thinks I'm slow." "I'm having the thought that this means I'm not good at my job." The linguistic awkwardness is the point.

Separate work from self. Before your next feedback conversation or performance review, remind yourself: the feedback will describe outputs I've produced, not who I am. Write it down if helpful. Create the mental separation in advance.

Ask for specific criticism proactively. This week, ask someone you trust: "What's one thing about my recent work that could be better?" Don't ask for general feedback—ask for specific criticism. Practice receiving it using the shifts. The more you practice in low-stakes situations, the better you'll handle high-stakes ones.

Troubleshooting

"Some criticism really is personal. What about when someone is actually attacking my character?"

It happens. Some people do make feedback personal. But here's the thing: even if they're attacking your character, you don't have to accept the framing. You can still choose to respond to the work component of their feedback and ignore the personal attack. "You seem to think I'm sloppy" / "Let me address the specific issues you mentioned in the document." You control which part of their message you engage with.

"I know intellectually that feedback is about the work, but I still feel terrible when I receive it. The shifts don't change the feeling."

The shifts don't eliminate the feeling. They change your relationship to the feeling. You can feel the sting of criticism while also recognizing that your interpretation is distorted. The feeling will fade faster when you're not feeding it with rumination and catastrophic thinking. Don't expect to feel nothing. Expect to recover faster.

"What if the criticism is simply wrong? Am I supposed to just accept inaccurate feedback?"

No. Sometimes criticism is wrong—factually inaccurate, based on misunderstanding, or reflecting the critic's limitations rather than yours. The shifts aren't about accepting all criticism as valid. They're about receiving criticism without defensiveness so you can evaluate it clearly. Once you've extracted yourself from the emotional spiral, you're in a much better position to assess whether the criticism has merit. Sometimes it does. Sometimes it doesn't. But you can't make that assessment while you're in defense mode.

• • •

Criticism is unavoidable in any professional environment. The question is not whether you'll receive it, but how you'll process it when it comes.

The shifts in this chapter give you a way through that doesn't require either defensive warfare or masochistic self-attack. You can receive feedback, extract its value, and move forward—without losing your equilibrium or your relationship with the critic.

In the next chapter, we tackle a related challenge: uncertainty and ambiguity. Criticism is clear—someone has identified a problem. But much of professional life involves operating when you don't know what's coming, when you can't be sure of the outcome, when the path forward is obscured. The cognitive skills for navigating uncertainty are different, and they're essential.

CHAPTER 5: Reframing Uncertainty and Ambiguity

The mental framework for operating when you can't see the outcome

> *Never let the future disturb you. You will meet it, if you have to, with the same weapons of reason which today arm you against the present.*
>
> — Marcus Aurelius, Meditations

You're considering a career move. The new role pays more, offers better growth potential, and aligns with your interests. But it's at a smaller company. What if they run out of funding? What if the culture is toxic? What if you hate it and can't get your old job back?

You don't know. You can't know. No amount of research will tell you what it's like to work there a year from now. No number of interviews will reveal the company's future or your satisfaction in a role you've never held.

So you hesitate. You request another call with the hiring manager. You ask more questions on Glassdoor. You make pro/con lists that somehow never resolve the uncertainty. You wait for a feeling of clarity that never arrives.

Meanwhile, the opportunity has a deadline. And your paralysis is itself a decision—a decision to stay where you are, made not by choice but by default.

This is how uncertainty wins. Not by being dangerous, but by being uncomfortable. You're not avoiding a bad outcome—you're avoiding the discomfort of not knowing.

• • •

The Certainty Illusion

Here's the uncomfortable truth: certainty doesn't exist.

You think your current job is stable. It isn't. Companies restructure, markets shift, managers leave, priorities change. You could be laid off next quarter for reasons entirely outside your control.

You think you know how a project will go. You don't. Requirements will change. Stakeholders will want different things. Unforeseen obstacles will emerge. The plan is a guess dressed up as a prediction.

You think you know what you want. Maybe. But you've also wanted things that turned out to be wrong for you. Your future self might have different priorities than your current self. You're uncertain even about your own preferences.

The feeling of certainty is not the presence of complete information. It's the absence of doubt—a psychological state that can exist regardless of actual knowledge. You can feel certain about things you're wrong about. You can feel uncertain about things that are clearly true.

When you wait for certainty before acting, you're waiting for a feeling, not a fact. And that feeling may never come—because the reality of most decisions is that you cannot know how they'll turn out until you make them.

The alternative isn't recklessness. It's competent action under uncertainty. Acting thoughtfully with incomplete information, knowing that incompleteness is the permanent condition of decision-making.

• • •

Analysis Paralysis and the Planning Fallacy

The brain has two contradictory impulses when facing uncertainty.

The first is to gather more information. If you're unsure, research more. Ask more questions. Run more analyses. The hope is that additional information will resolve the uncertainty and make the right choice obvious.

This works up to a point. The first wave of research usually adds significant value. The second wave adds less. By the third and fourth waves, you're in diminishing returns—learning marginally more while delaying indefinitely.

This is analysis paralysis. The information-gathering becomes an end in itself, a way to feel productive while avoiding the discomfort of deciding. You're not actually getting closer to a decision. You're postponing it.

The second impulse is the planning fallacy—the tendency to underestimate how long things will take, how complex they'll be, and how many obstacles will emerge. When you finally do act, you assume a best-case scenario: everything goes smoothly, nothing unexpected happens, the plan unfolds as written.

These two impulses create a painful combination. You delay action because you're uncertain, then act as if you're certain. You research obsessively to reduce risk, then ignore risk in your execution.

The solution isn't to plan less or research more. It's to accept that uncertainty is irreducible. You will never have complete information. Your plans will never survive contact with reality intact. The question is not how to eliminate uncertainty but how to act effectively within it.

• • •

Catastrophizing: The "What If" Spiral

When uncertainty meets anxiety, the mind generates worst-case scenarios.

What if the new job doesn't work out? What if you fail publicly? What if you make the wrong choice and ruin everything?

This is catastrophizing—jumping to the worst possible outcome and treating it as likely. The mind doesn't just acknowledge that things could go wrong; it elaborates the disaster in vivid detail. You don't just worry about failure; you imagine the humiliation, the consequences, the cascade of problems that would follow.

Catastrophizing feels like preparation. If you can anticipate the worst, maybe you can prevent it. But catastrophizing isn't preparation—it's rumination. You're not planning responses to possible problems. You're marinating in imagined disasters.

The cognitive pattern typically follows a chain. Start with any uncertainty, and the catastrophizing mind can connect it to ruin in three or four steps:

- What if this presentation doesn't go well?
- Then my reputation will be damaged.
- Then I won't get promoted.
- Then my career will stall.
- Then I'll be stuck and miserable forever.

Each step feels logical, but the chain as a whole is absurd. The probability of all those things happening in sequence is tiny. But the mind doesn't compute joint probabilities. It follows the chain of worry link by link, each step feeling inevitable.

Breaking the chain requires interrupting the "what if" with "what then." Not "what if this goes wrong" but "what will I do if this goes wrong." The first question spirals into catastrophe. The second generates a plan.

• • •

Premeditatio Malorum: The Stoic Premeditation of Adversity

The Stoics had a practice that looks superficially like catastrophizing but functions completely differently. They called it *premeditatio malorum*—the premeditation of adversity.

Seneca described it this way: "What is quite unlooked for is more crushing in its effect, and unexpectedness adds to the weight of a disaster... Nothing ought to be unexpected by us. Our minds should be sent forward in advance to meet all problems, and we should consider, not what is wont to happen, but what can happen."

The practice involves deliberately imagining worst-case scenarios—but not to wallow in them. The purpose is to prepare psychologically and practically.

When you imagine what could go wrong before it happens, two things occur. First, the scenario loses some of its power. You've already faced it mentally, so it's less shocking if it occurs. You're not blindsided. Second, you can develop responses in advance. If this happens, I'll do that. The contingency plan is ready.

This is the opposite of catastrophizing. Catastrophizing asks "what if?" and spirals into fear. Premeditatio asks "what if?" and plans for action.

The key difference is agency. Catastrophizing treats bad outcomes as things that happen to you. Premeditatio treats them as things you can respond to. The first is passive terror. The second is active preparation.

Practicing premeditatio malorum might look like this: Before launching a project, ask "What could go wrong?" List the realistic possibilities. For each one, ask "What would I do if this happened?" Develop contingencies. Then proceed with the project, knowing you've prepared for adversity.

You haven't eliminated uncertainty. You've reduced your vulnerability to it.

• • •

ACT and Willingness to Move Toward Discomfort

Acceptance and Commitment Therapy offers another lens on uncertainty: the concept of willingness.

Willingness isn't wanting something to happen. It's openness to experiencing discomfort in service of your values. You might not want to feel the anxiety of uncertainty. But you can be willing to feel it—because avoiding that feeling would mean abandoning goals that matter to you.

The ACT framework distinguishes between "clean" discomfort and "dirty" discomfort. Clean discomfort is the pain inherent in a situation—the unavoidable anxiety of making a high-stakes decision, the nervousness of starting something new, the fear of potential failure. Dirty discomfort is the suffering you add through resistance, avoidance, and rumination.

Uncertainty produces clean discomfort. You don't know how things will turn out. That's uncomfortable. It's supposed to be uncomfortable. Your brain is signaling that something unknown is ahead.

The dirty discomfort comes from fighting the uncertainty. The weeks of indecision. The obsessive information-gathering. The mental spiraling through worst-case scenarios. The avoidance of decisions that feel too risky. This isn't uncertainty causing suffering—it's your resistance to uncertainty causing suffering.

Willingness means accepting the clean discomfort so you don't create the dirty kind. Yes, this decision feels uncomfortable. Yes, I can't be certain. Yes, things might go wrong. I'm willing to feel all of that—because the alternative is paralysis.

Moving toward discomfort rather than away from it is counterintuitive. The instinct is to reduce discomfort. But avoidance often increases suffering in the long run. You don't get relief from uncertainty by avoiding decisions—you get prolonged anxiety and missed opportunities.

The willing stance says: I don't need certainty to act. I can tolerate not knowing. Let's go.

• • •

The Four Uncertainty Shifts

SHIFT 15

FROM: *"I can't act until I'm certain."*

TO: *"Certainty is a fantasy; competent action isn't."*

This shift attacks the core assumption that certainty is achievable and necessary.

"I can't act until I'm certain" sounds reasonable. It seems prudent to wait until you know enough. But it sets an impossible standard. You will never be certain. There will always be unknown variables, unpredictable factors, information you don't have.

If certainty is the prerequisite, action becomes impossible.

"Certainty is a fantasy" isn't cynicism—it's realism. It acknowledges the actual conditions under which all decisions are made. Nobody knows how things will turn out. Successful people aren't successful because they had certainty. They're successful because they acted competently despite uncertainty.

"Competent action isn't" is the crucial addition. You can't achieve certainty, but you can achieve competence. You can gather reasonable information. You can think through implications. You can prepare for likely contingencies. You can make the best decision available given what you know.

Competent action under uncertainty is a skill. It can be developed. And it produces better outcomes than waiting for certainty that never arrives.

SHIFT 16

> **FROM:** *"What if this goes wrong?"*
>
> **TO:** *"What will I do if this goes wrong?"*

This shift transforms catastrophizing into preparation.

"What if this goes wrong?" is a question with no useful answer. It invites imagination without direction. The mind generates disasters. Anxiety escalates. No action follows.

"What will I do if this goes wrong?" demands an answer. It's not asking whether bad things could happen—of course they could. It's asking how you'll respond when they do.

The first question is about the world. What might happen to me? The second question is about you. What will I do?

This is premeditatio malorum in action. You're not avoiding the possibility of failure. You're planning for it. If the project fails, I'll extract the lessons and try a different approach. If the job doesn't work out, I'll start looking within six months and use my new skills elsewhere. If the launch flops, I'll analyze what went wrong and iterate.

Each contingency reduces the weight of the uncertainty. You've already faced the worst case mentally. You have a plan. The scenario doesn't disappear—but it stops spiraling.

SHIFT 16

FROM: *"I need more information."*

TO: *"I need to decide with what I have."*

This shift breaks analysis paralysis.

"I need more information" feels responsible. Who could argue with gathering more data? But at some point, the information-gathering becomes avoidance. You're not researching to decide better—you're researching to postpone deciding.

"I need to decide with what I have" accepts the constraint. The information available is the information available. Waiting won't change that significantly. The uncertainty that remains will remain.

This doesn't mean ignoring useful information. The first round of research usually matters. The question is whether additional research is actually adding value or just delaying.

A useful test: If you gathered the information you're seeking, would it change your decision? If the answer is "probably not," you don't need more information. You need to decide.

Amazon's Jeff Bezos talks about "70% decisions"—acting when you have about 70% of the information you wish you had. Waiting for 90% information costs speed. And in fast-moving environments, the cost of slowness exceeds the benefit of additional certainty.

Most decisions can be made earlier than you think. Most information-gathering past the initial phase is delay in disguise.

SHIFT 16

FROM: *"The unknown is threatening."*

TO: *"The unknown is where opportunity lives."*

This shift reframes uncertainty itself.

Your brain treats the unknown as dangerous. That's the threat-detection system doing its job. Unknown territory might contain predators. Unknown situations might cause harm. Better to stay with the familiar.

But the unknown is also where everything new happens. Every opportunity you've ever seized required moving into uncertainty. The job offer, the new relationship, the creative project, the business venture—none of these came with guarantees. They required tolerating not-knowing.

If you only go where outcomes are certain, you go nowhere new.

"The unknown is where opportunity lives" isn't naive optimism. It doesn't claim that all uncertainty leads to good outcomes. It claims that good outcomes require engaging with uncertainty. The path to growth runs through discomfort.

This reframe doesn't eliminate the threat response. Your body will still tighten when facing the unknown. But you can hold a different cognitive frame while that happens. Yes, this is uncertain. Yes, that feels uncomfortable. And also: this is where new possibilities exist. The discomfort is the price of admission.

People who achieve remarkable things are not people who somehow escaped uncertainty. They're people who developed a different relationship to it. They learned to see the unknown not as something to avoid but as territory to explore.

• • •

Putting It Together

When facing uncertainty, you have a practice available.

Notice the paralysis. Are you researching past the point of usefulness? Are you waiting for clarity that won't come? Are you avoiding a decision because it feels risky?

Identify the cognitive pattern. Is this catastrophizing—spiraling through worst-case scenarios? Is this the certainty demand—requiring information that doesn't exist? Is this analysis paralysis—research as avoidance?

Apply the relevant shift. Transform "what if" into "what will I do." Accept that certainty is a fantasy. Decide with what you have. Recognize that opportunity lives in uncertainty.

Practice premeditatio malorum. If the worst happened, what would you do? Develop contingencies. Face the scenario mentally. Reduce its power through preparation.

Embrace willingness. Accept the clean discomfort of uncertainty. Choose action over the dirty discomfort of endless avoidance.

Act. Not recklessly. Competently. With the information available. Knowing you might be wrong. Proceeding anyway.

• • •

TAKE ACTION: YOUR NEXT STEPS

Reflection

Identify a decision you've been postponing due to uncertainty. Work through these questions:

1. What certainty are you waiting for? What information or assurance do you feel you need before acting? Be specific.

2. Is that certainty achievable? Honestly assess: Can you actually get the information you're waiting for? Or are you waiting for something that doesn't exist?

3. What's the cost of delay? What are you losing by not deciding? Opportunity cost? Time? Energy spent ruminating? Other options closing?

4. What would you do if the worst case happened? Apply premeditatio malorum. If things went wrong in the way you fear, what would your response be? What's your contingency plan?

5. What does competent action look like here? Given the information you have, what's the thoughtful, prepared decision? Not the certain decision—the competent one.

Implementation

Set decision deadlines. For decisions you've been postponing, set a date by which you will decide. Not "when I have enough information" but a specific date. This forces you to shift from gathering to deciding.

Practice the "what will I do" reframe. This week, catch yourself catastrophizing. Every time you notice "what if this goes wrong," convert it to "what will I do if this goes wrong." Answer the question. Make a contingency plan, even if it's rough.

Take one "70% decision." Identify something you've been researching extensively. Ask yourself: Do I have 70% of the useful information? If yes, decide. Notice what happens. Most decisions are reversible or adjustable. You can correct course as you learn more.

Troubleshooting

"But some decisions really are high-stakes. Shouldn't I be more careful with those?"

High stakes warrant appropriate diligence—not infinite delay. The question is whether additional research actually reduces risk or just postpones decision. For truly high-stakes choices, you can do scenario planning, seek expert input, and develop robust contingencies. But eventually you still have to decide with incomplete information. The stakes don't change that.

"I've made decisions under uncertainty before and they went badly. Isn't caution justified?"

Bad outcomes don't prove the decision was wrong. Uncertainty means things sometimes go badly even when the decision was sound. The question is whether you made a competent decision given what you knew at the time—not whether the outcome was good. If your process was solid, a bad outcome is data for next time, not evidence you should never decide under uncertainty again.

"What if my gut is telling me something is off? Shouldn't I listen to that?"

Intuition is real and sometimes valuable. But intuition is also shaped by anxiety, bias, and past experience that may not apply to the current situation. The question is whether your gut is sensing something real about this situation or just signaling general discomfort with uncertainty. If you can articulate what specifically feels wrong, investigate that. If it's just a vague sense of unease—that's probably your brain protesting the unfamiliar. That's not a signal to retreat. That's a signal you're facing growth.

• • •

Uncertainty is the permanent condition of professional life. Markets shift, technologies change, people surprise you. The fantasy of certainty is just that—a fantasy. The people who thrive aren't those who somehow achieved certainty. They're those who learned to act effectively without it.

The shifts in this chapter give you a different relationship with the unknown. Not comfort—uncertainty doesn't become comfortable. But competence. The ability to move forward when the path isn't clear. The willingness to tolerate not knowing in service of values that matter.

In the next chapter, we turn to a different challenge: status and comparison. When you look at others in your field—the people who seem ahead, who seem to have what you want, who seem to have figured it out—what happens in your mind? For many, it's a source of quiet suffering. The cognitive shifts that follow address how to pursue excellence without the psychological tax of constant measurement against others.

CHAPTER 6: Reframing Status and Comparison

How to pursue excellence without the psychological tax of constant measurement

> *How much time he gains who does not look to see what his neighbor says or does or thinks, but only at what he himself is doing, to make it just and holy.*
>
> — Marcus Aurelius, Meditations

You open LinkedIn. A former colleague just got promoted to VP. Someone you graduated with raised $20 million for their startup. A peer from your first job is now a keynote speaker at industry conferences.

What happens in your body? For most people, something deflates. A quiet constriction. The internal monologue starts: Why not me? What am I doing wrong? I should be further along by now.

You didn't choose to feel this way. You didn't decide to compare yourself to these people. The comparison happened automatically, and it left you feeling inadequate—even though five minutes ago, before you opened the app, you felt fine about your work and your life.

This is the comparison trap. It operates constantly, mostly below conscious awareness. And it extracts a psychological tax that compounds over time—draining energy, distorting priorities, and making achievement feel hollow even when it comes.

This chapter is about breaking free from that trap. Not by avoiding all information about others' success—that's impossible. But by developing a different relationship to that information.

• • •

Social Comparison Theory and Its Costs

In 1954, psychologist Leon Festinger proposed social comparison theory. The basic idea: humans have a drive to evaluate themselves, and in the absence of objective standards, we evaluate by comparing to others.

This was probably adaptive in small tribal groups. Knowing your relative standing helped you navigate social hierarchies, identify learning opportunities, and motivate improvement. If the hunter next to you was catching more game, that was useful information.

But the modern environment has broken this system.

You're no longer comparing yourself to fifteen people in your village. You're comparing yourself to billions of people whose highlight reels appear in your pocket throughout the day. The hunter next to you isn't just the guy in the neighboring hut—it's every successful person whose achievements are algorithmically surfaced to grab your attention.

The math doesn't work. There will always be someone with more money, more status, more followers, more accolades. No matter how much you achieve, the comparison pool contains people who achieved more. The treadmill has no endpoint.

Research on social comparison shows consistent findings. Upward comparison—comparing to those doing better—tends to produce negative emotions: envy, inadequacy, reduced self-esteem. These feelings occur even when the comparison is illogical, even when you know the other person had advantages you didn't, even when their success has no bearing on yours.

The costs compound. Time spent ruminating on others' achievements is time not spent on your own work. Mental energy devoted to feeling inadequate is energy unavailable for creating. The emotional drain of constant comparison reduces both performance and satisfaction.

And here's the cruelest part: achieving more doesn't solve it. When you reach a new level, you start comparing to people at that level. The promotion you desperately wanted feels underwhelming once you have it, because now you're measuring against people with bigger promotions. The treadmill speeds up.

• • •

The "Should" Distortion

Comparison often expresses itself through "should" statements.

I should be making more money by now. I should have been promoted already. I should be further along in my career. I should have what they have.

These "shoulds" feel like reasonable standards. They seem to represent legitimate expectations. But examine them closely and they dissolve.

Where does the "should" come from? Who decided you should be making more money by this age? What law establishes when you should be promoted? The standards aren't objective—they're internalized comparisons dressed up as rules.

"I should be further along" is actually "People I'm comparing myself to are further along, and I've decided their trajectory should be mine." But why should their trajectory be yours? They had different starting points, different advantages, different priorities, different trade-offs. The comparison ignores all of that.

"Should" statements create suffering by positing an alternative reality that doesn't exist. You should be further along—but you're not. Now what? The "should" doesn't help you move forward. It just makes you feel bad about where you are.

The cognitive distortion here is emotional reasoning combined with arbitrary standards. You feel like you should be further along, so you assume there's a legitimate basis for that feeling. There isn't. The feeling comes from comparison, and the comparison is arbitrary.

Dropping the "should" doesn't mean abandoning ambition. You can want to advance without believing you should already be advanced. You can work toward goals without treating your current position as a failure.

• • •

Values Clarification vs. External Metrics

Comparison uses external metrics: titles, salaries, followers, awards. These metrics are visible, quantifiable, and easy to rank. That's precisely why they're dangerous.

External metrics measure outcomes that are partly outside your control. You can do excellent work and not get the promotion because of politics, timing, or budget. You can build a great product and not get the funding because of market conditions or investor preferences. External metrics reflect your efforts, but they also reflect luck, circumstance, and factors that have nothing to do with you.

When you define success by external metrics, you hand your sense of achievement to forces you can't control. You might do everything right and still "fail" by those measures. You might do things you're not proud of and "succeed." The metric and the meaning come apart.

Values clarification offers an alternative. Instead of asking "How do I compare to others?" you ask "Am I living according to what I believe matters?"

This isn't about ignoring achievement or pretending you don't want success. It's about grounding your evaluation in something you actually control: your own choices, integrity, and effort.

What do you value? If you value learning, are you learning? If you value contribution, are you contributing? If you value integrity, are you acting with integrity? If you value presence, are you present with the people who matter?

These questions don't require comparison to anyone else. They require honest self-assessment against standards you've chosen. And they produce a different kind of satisfaction—one that doesn't depend on where you rank in a status hierarchy.

ACT calls this "values-driven action." The good life isn't defined by achievements accumulated but by alignment between what you do and what you believe matters. That alignment is available at any stage of career or life. It doesn't require being ahead of anyone.

• • •

The Stoic Focus on Character Over Circumstance

The Stoics had a radical position on external goods: they're "preferred indifferents."

Wealth, status, reputation, health—these things are preferred in the sense that most people would rather have them than not. But they're indifferent in the sense that they don't determine whether you're living a good life. You can have all of them and be miserable. You can lack them and flourish.

What determines a good life? Character. Virtue. Living according to wisdom, justice, courage, and temperance. These are the only true goods because they're the only things fully within your control.

This framework liberates you from comparison. If the good life depends on external achievement, you're forever racing against others who might be ahead. If the good life depends on character, the only relevant question is whether you're living with integrity—and that's a question you answer through your own choices.

Marcus Aurelius was emperor of Rome. He had more power, wealth, and status than almost anyone in history. And he wrote to himself, repeatedly, not to care about these things. "Never value anything as profitable that compels you to break your promise, lose your self-respect, hate any man, suspect, curse, act the hypocrite, or desire anything that needs walls or curtains."

He understood that external success, pursued at the cost of character, wasn't success at all. And external success, achieved while maintaining integrity, wasn't the point—the integrity was.

This doesn't mean abandoning ambition. Marcus was arguably the most powerful man in the world. The Stoics weren't passive. But they held achievement loosely, as a preferred indifferent rather than the measure of a life.

You can work hard, build things, seek advancement—while remembering that these achievements don't define your worth. Your worth is defined by how you choose to live, not by where you rank.

• • •

The Four Comparison Shifts

SHIFT 19

FROM: *"They're ahead of me."*

TO: *"We're running different races."*

This shift challenges the assumption that you and the person you're comparing to are on the same track.

"They're ahead of me" implies a single race with a clear leader. But careers aren't like that. People have different goals, different values, different constraints, different definitions of success.

Your colleague who made VP might be optimizing for corporate advancement. You might be optimizing for work-life balance, creative autonomy, or skill development. These are different races. Comparing positions is meaningless when you're not going to the same destination.

The person who raised $20 million might have spent a decade in grinding poverty while building their startup. You might have chosen steady income and time with family. Neither is wrong. They're different paths with different trade-offs.

"We're running different races" doesn't deny that some people have achieved more by certain metrics. It denies that those metrics are the universal scorecard. They're one possible scorecard—and you might not be playing that game.

The shift creates space for appreciation without inadequacy. You can acknowledge someone's achievements without concluding you're failing. You can learn from their path without judging yours. The races are different. There's no leaderboard that matters.

SHIFT 20

FROM: *"I should be further along."*

TO: *"I'm where my decisions brought me; now I decide again."*

This shift replaces the "should" distortion with radical acceptance and agency.

"I should be further along" posits an alternative timeline where you made different choices and got different results. But that timeline doesn't exist. The only timeline that exists is the one where you made the choices you made.

Those choices brought you here. Some were good, some were mistakes, some were constrained by circumstances outside your control. The result is your current position. That's reality.

"I'm where my decisions brought me" is acceptance without self-attack. You're not lamenting your position or beating yourself up. You're acknowledging the causal chain that led here. The past is past.

"Now I decide again" is where agency returns. You can't change where you are. You can change what you do next. The question isn't "Why aren't I further along?" but "What decisions can I make from here?"

This shift is liberating because it ends the argument with the past. The "should" keeps you stuck, relitigating choices that can't be unmade. The shift moves you forward: Here's where I am. Now what?

SHIFT 20

FROM: *"Success means having what they have."*

TO: *"Success means living according to my values."*

This shift relocates the definition of success from external to internal.

"Success means having what they have" outsources your life's meaning to someone else's achievements. You're not pursuing your vision—you're chasing their results. Even if you get them, you might find they don't fit you. The VP role that works for them might drain you. The startup that thrills them might exhaust you.

Worse, the target keeps moving. Once you have what they have, you'll compare to someone who has more. There's no arrival.

"Success means living according to my values" grounds success in something stable: your own chosen principles. If you value depth over breadth, success is going deep—regardless of what others are doing. If you value impact over income, success is making impact. If you value autonomy over advancement, success is being your own boss, even if that means a smaller operation.

Values-based success isn't easier to achieve, but it's clearer to define. You know what you're pursuing. And you can assess your progress without reference to anyone else.

This doesn't require ignoring external achievement entirely. Values and ambition aren't mutually exclusive. But the values come first. Achievement serves the values, not the other way around.

SHIFT 22

FROM: *"I'm behind."*

TO: *"I'm on my timeline."*

This shift rejects the premise that there's a standard timeline against which you're measured.

"I'm behind" assumes a race everyone is running together, with a pace you're supposed to match. But who set that pace? The expectation that you should hit certain milestones by certain ages is arbitrary—a cultural construction, not a natural law.

Some people advance quickly early and plateau. Some people start slow and accelerate later. Some take detours that look like setbacks but create unique capabilities. Some follow straight paths that look efficient but feel hollow.

There is no universal timeline. There's only your timeline.

"I'm on my timeline" isn't resignation or excuse-making. It's acknowledgment that your path is yours—shaped by your choices, circumstances, and priorities. It unfolds at its own pace.

This shift neutralizes one of comparison's main weapons: the sense that you're falling behind. You can't fall behind in a race you're not running. You're not late to someone else's milestones. You're on your own schedule.

The practical implication: instead of measuring against external timelines, measure against your own trajectory. Are you moving in a direction you value? Are you growing? Are you doing better work than you did a year ago? These questions matter. Where you rank against peers does not.

• • •

Putting It Together

When comparison strikes, you have a sequence available.

Notice the trigger. What just activated the comparison? A LinkedIn post? A conversation? A status update? Name it.

Catch the interpretation. What story did your mind generate? "They're ahead." "I'm behind." "I should be further." Name the specific thought.

Apply the relevant shift. We're running different races. I'm where my decisions brought me. Success is living my values. I'm on my timeline.

Reconnect to your values. What do you actually want from your life and career? Not what you're supposed to want. Not what would impress others. What matters to you?

Redirect to action. What can you do right now, from where you are, in service of what you value? That's the only productive question.

This won't eliminate comparison forever. The instinct is deep. But you can reduce its grip. Each time you catch a comparison and reframe it, you weaken the automatic pattern. Over time, you build a different relationship to others' success—one where you can appreciate it, learn from it, even be inspired by it, without it diminishing you.

• • •

TAKE ACTION: YOUR NEXT STEPS

Reflection

Think of a comparison that's been bothering you—someone whose success makes you feel inadequate. Work through these questions:

1. **What specifically are you comparing?** Be precise. Is it title? Income? Recognition? Lifestyle? Name the metric you're using.

2. What do you actually know about their situation? What trade-offs did they make? What advantages did they have? What costs have they paid? You probably know less than you think.

3. What "should" statement is operating? Complete the sentence: "I should be..." Where did that expectation come from? Is it your standard or something you absorbed?

4. What do you value? If you clarified your own values, what would success look like? Does it match what you're envying in this person?

5. What race are you actually running? If you defined your path by your own priorities, would this comparison even be relevant?

Implementation

Conduct a comparison audit. For one week, track every time you compare yourself negatively to someone else. Note: what triggered it, what thought arose, and how you felt. At the end of the week, review the patterns. What triggers you most? What thoughts recur?

Write your values definition of success. Take thirty minutes to write out what success means to you—not what you're supposed to want, not what would impress your parents or peers. Your definition. Be honest. Use this as a touchstone when comparison strikes.

Practice celebrating others' success. The next time you see someone achieve something that triggers comparison, deliberately reframe. "Good for them. We're running different races. Now, what am I working on?" Turn the comparison into a reminder of your own path.

Troubleshooting

"But comparison is sometimes motivating. Shouldn't I use it as fuel?"

Comparison can motivate, but it's expensive fuel. The motivation comes with anxiety, inadequacy, and a destination defined by someone else. Values-based motivation is cleaner—you pursue what matters to you because it matters, not because you're trying to catch up to someone. If comparison motivates you, ask: at what cost? And could you find a less toxic source of drive?

"Some comparisons are legitimate. If I'm underperforming relative to peers, don't I need to know that?"

Performance relative to expectations is worth understanding. If you're falling short of what your role requires, that's useful data. But that's different from status comparison—measuring your career against others' highlight reels. One is functional feedback. The other is psychological self-harm.

"I can't just stop comparing. It's automatic."

You're right—you can't eliminate comparison through willpower. The goal isn't to never compare but to change your response when comparison happens. Catch it. Name it. Reframe it. Over time, the automatic pattern weakens. You won't stop noticing others' success. You'll stop letting it determine your self-worth.

• • •

Comparison is a trap because it has no exit. There will always be someone ahead. The treadmill never stops. The only way to win is to step off—to define success by your own values rather than your rank on someone else's scoreboard.

The shifts in this chapter give you a way to do that. Not by ignoring achievement or pretending you don't want success. But by locating success where it belongs: in how you live, not how you compare.

In the final career chapter, we address the question of meaning itself. Beyond the metrics, beyond the comparisons, beyond the achievements—what makes work feel worthwhile? The cognitive shifts that follow help you find purpose in contribution rather than recognition, in the work itself rather than its rewards.

CHAPTER 7: Reframing Work and Contribution

The shifts that reconnect effort to meaning

> *Waste no more time arguing about what a good man should be. Be one.*
>
> — Marcus Aurelius, Meditations

You've been at this job for three years. The initial excitement faded long ago. Now it's just work—meetings, emails, deliverables, deadlines. The same problems recycling. The same conversations repeating. Nothing you do seems to matter in any larger sense.

You look at your calendar and feel a flatness. Another day of tasks. Another week of incremental progress on projects that feel disconnected from anything meaningful. You're earning a paycheck. You're building a resume. But somewhere along the way, the sense of purpose leaked out.

This isn't burnout exactly. You're not exhausted. You're not overwhelmed. You're just... empty. The work gets done, but it doesn't feed you.

This chapter is about recovering meaning. Not by changing jobs or chasing passion—though those might be right for some people. But by changing how you see the work you're already doing. Sometimes the meaning was there all along, hidden behind cognitive filters you didn't know you were wearing.

• • •

The Mental Filter That Ignores Contribution

Your brain has a filter for what counts as meaningful. And that filter is probably miscalibrated.

When you imagine "work that matters," what comes to mind? Probably big, visible achievements. Launching a product that changes an industry. Leading a team that breaks records. Getting public recognition for innovation or impact. These are the images of meaningful work that culture celebrates.

By comparison, your actual work looks small. You answered emails. You fixed a bug. You helped a colleague solve a problem. You made a spreadsheet slightly better. You clarified requirements for a project. None of this registers as mattering.

But this is a filtering error. You're discounting your contribution because it doesn't match the culturally celebrated image of significance.

Consider what actually happens when you help that colleague. They're stuck on a problem. It's blocking their progress. Their stress is rising. You spend fifteen minutes explaining something, and suddenly they're unstuck. They can move forward. Their day improves. The project advances.

That's contribution. That matters. Not in a world-changing way—but in a real way to a real person dealing with a real problem.

The filter ignores this because it's looking for grand gestures. It misses the texture of daily contribution—the countless small ways you make things better for people around you. Over time, this filtering error creates the experience of meaninglessness even when you're contributing constantly.

• • •

Discounting the Positive

There's a specific cognitive distortion at work here: discounting the positive.

Discounting the positive means dismissing good things as not counting. You helped someone, but "anyone could have done that." You completed a project successfully, but "it wasn't that hard." You received positive feedback, but "they're just being nice."

This distortion systematically strips meaning from your experience. No matter what you contribute, the discounting filter explains it away. The result is a record of achievements that somehow doesn't register as achievement. You've done things, but they don't count.

The distortion operates through comparison and minimization. Whatever you did, someone else did more. Whatever you achieved, it could have been bigger. The bar for "counting" keeps rising, staying perpetually out of reach.

Notice the asymmetry. When something goes wrong, it counts completely. The mistake, the failure, the criticism—these register fully. But when something goes right, it gets discounted. The successes don't count; the failures do. No wonder work feels meaningless.

Breaking this pattern requires deliberately crediting your contributions. Not inflating them—just accurately recognizing them. You helped someone. That counted. You solved a problem. That counted. You made something better than it was before. That counted.

• • •

Values-Driven Action vs. Avoidance-Driven Action

ACT distinguishes between two kinds of motivation: moving toward what you value and moving away from what you fear.

Avoidance-driven action is about escaping discomfort. You work to avoid getting fired. You complete tasks to avoid criticism. You show up to avoid disappointing people. The engine is fear—fear of negative consequences if you don't do what's expected.

Avoidance-driven work is exhausting because there's no positive destination. You're running from something, not toward something. When the fear temporarily recedes, there's nothing left to motivate you. The work feels like a treadmill of threat management.

Values-driven action is about moving toward what matters. You work because the work itself serves something you care about. You complete tasks because they contribute to outcomes you value. You show up because presence aligns with who you want to be.

Values-driven work can involve the same tasks as avoidance-driven work. The difference is internal. You're not fleeing pain—you're pursuing meaning. And that pursuit doesn't exhaust in the same way. It generates energy rather than draining it.

The shift from avoidance to values isn't about changing what you do. It's about connecting what you do to why it matters. The email you're dreading isn't just an obligation to discharge—it's communication that moves a project forward. The meeting you're avoiding isn't just a time sink—it's a chance to align people toward a shared goal.

This reframing doesn't make difficult work easy. But it changes the relationship to that work. You're not just surviving. You're contributing.

• • •

Marcus Aurelius on Doing Good Work Quietly

Marcus Aurelius spent much of his journal reminding himself not to seek recognition.

"Do not act as if you had ten thousand years to live... while you have life in you, while you still can, make yourself good."

"If you do the job in a principled way, with diligence, energy and patience, if you keep yourself free of distractions... if you can embrace this without fear or expectation... you can find fulfillment."

"It is not death that a man should fear, but he should fear never beginning to live."

The emperor of Rome—a man who could have had any recognition he wanted—was counseling himself to focus on the work itself rather than the applause. Why?

Because recognition is outside your control. You can do excellent work and be ignored. You can do mediocre work and be celebrated. The correlation between quality and recognition is imperfect at best. If you need recognition to feel your work matters, you've handed meaning to forces you can't influence.

The work itself is within your control. You can choose to do it well. You can choose to invest care and skill. You can choose to make something better than it needs to be. That choice doesn't depend on whether anyone notices.

Marcus called this "doing good work quietly"—contributing not for the audience but for the contribution itself. The meaning comes from the doing, not from the response to the doing.

This isn't about rejecting recognition when it comes. Recognition is a preferred indifferent—nice to have, but not the point. The point is the work. Do it well. Let the recognition take care of itself.

• • •

The Compound Effect of Small Contributions

Individual contributions feel small because they are small—in isolation. But contributions don't exist in isolation. They compound.

The fifteen minutes you spent helping a colleague today connects to the hour they'll save tomorrow, which connects to the project finishing on time, which connects to the client being served, which connects to the business thriving, which connects to jobs and livelihoods and value created in the world.

You can't see these connections in the moment. The feedback loop is too long. But the connections are real.

Think of your work as a single brick. A brick by itself is just a rectangle of clay. It's not impressive. It doesn't look like much. But bricks compound into walls, and walls compound into buildings, and buildings compound into cities. The brick doesn't become meaningful by being bigger. It becomes meaningful by being part of something bigger.

Your daily contributions are bricks. They don't look like much in isolation. But they're building something. The question is whether you can see past the single brick to the structure emerging.

This perspective doesn't require believing your current work is your life's calling. You might be in a role that's temporary, a stepping stone to something else. Even then, the contributions compound—skills built, relationships formed, problems solved. The brick goes into some structure, even if it's not the final one.

• • •

The Four Meaning Shifts

SHIFT 23

FROM: *"This work doesn't matter."*

TO: *"Every contribution compounds."*

This shift attacks the filtering error that ignores small contributions.

"This work doesn't matter" applies a standard of grand impact that most work can't meet. By that standard, almost nothing matters—and that's a useless framework for finding meaning in daily effort.

"Every contribution compounds" recognizes that significance accumulates. The email you sent moved something forward. The problem you solved prevented a downstream failure. The colleague you helped became slightly more effective. None of these is world-changing. All of them are real.

Compound effects are hard to see because they unfold over time and across connections you can't trace. But that doesn't mean they're not happening. Every action ripples outward. Every contribution becomes part of a larger pattern.

The shift invites you to trust the compounding even when you can't see it. Your work matters not because each piece is independently significant but because each piece contributes to something larger. The brick doesn't need to be impressive. It needs to be solid and placed well.

SHIFT 24

FROM: *"I'm just doing a job."*

TO: *"I'm solving problems for people."*

This shift reframes work as service.

"I'm just doing a job" reduces work to transactional exchange—labor for money. The meaning drains out because the framing excludes meaning. It's just a job. Just tasks. Just obligations.

But behind every job are people with problems. The software you build helps someone accomplish something they couldn't accomplish before. The reports you create inform decisions that affect real outcomes. The support you provide reduces someone's burden or confusion.

"I'm solving problems for people" reconnects the tasks to their purpose. Not abstract purpose—concrete people with concrete needs that your work addresses.

This isn't about inflating importance. Not every problem you solve is life-or-death. But problems don't need to be dramatic to matter to the people experiencing them. The user who couldn't figure out the interface, the client who needed data to make a decision, the colleague who was stuck and is now unstuck—these are real people you helped.

The shift works even for work that feels deeply unglamorous. Someone has to process the invoices. If they don't, people don't get paid. Someone has to maintain the database. If they don't, the system fails. The work is mundane, but the problems it solves are real.

SHIFT 22

FROM: *"I deserve more recognition."*

TO: *"I choose work worth doing regardless of applause."*

This shift addresses the recognition trap.

"I deserve more recognition" might be true. Many people are under-recognized. Their contributions go unnoticed. Their work is taken for granted. The frustration is legitimate.

But the statement contains a hidden dependency: your sense of meaning requires external validation. If you deserve recognition and don't get it, you feel cheated. The work loses its meaning because the expected reward didn't arrive.

"I choose work worth doing regardless of applause" reclaims agency. You're not doing the work for recognition. You're doing it because it's worth doing—because it aligns with your values, serves people, builds something, solves problems. The applause is secondary.

This isn't about suppressing the desire for recognition. That desire is human and valid. But desire and dependency are different. You can want recognition without needing it to feel your work matters.

The shift aligns with Marcus Aurelius's practice. Do good work quietly. Let the meaning come from the doing. If recognition follows, fine. If it doesn't, you still did work worth doing.

SHIFT 26

> **FROM:** *"I'm stuck here."*
>
> **TO:** *"I'm building capacity here."*

This shift reframes constraint as opportunity.

"I'm stuck here" treats your current position as a trap. You can't move. You're confined. The situation is static and limiting.

But even if you're in a role you'd rather leave—even if you're working toward something else—the current position isn't wasted time. It's building capacity.

Every challenge you navigate builds problem-solving skill. Every difficult conversation builds communication ability. Every boring task done well builds discipline. Every constraint you operate within builds resourcefulness.

"I'm building capacity here" doesn't require loving your current job. It requires recognizing that the job is developing you even when it's frustrating. The skills you build are portable. The experience accumulates. The capacity grows.

This perspective transforms the feeling of being stuck. You're not trapped—you're in training. The current role might not be the destination, but it's building the capabilities you'll need for wherever you're going.

• • •

Putting It Together

When work feels meaningless, you have a sequence available.

Notice the emptiness. Don't dismiss the feeling. It's information. Something in your framing has disconnected effort from meaning.

Identify the filter. What's your mind discounting? Are you ignoring small contributions because they're not grand enough? Are you dismissing achievements because they don't meet some arbitrary threshold? Are you missing the people your work actually serves?

Apply the relevant shift. Every contribution compounds. You're solving problems for people. You choose work worth doing regardless of applause. You're building capacity.

Reconnect to values. What do you value that this work serves? Even partially? Even imperfectly? The connection might be subtle, but it's probably there.

Credit your contribution. Deliberately. Today you helped someone. Today you made something work. Today you solved a problem. Let that count.

This won't transform meaningless work into passionate calling. But it can close the gap between what you're doing and why it matters. The meaning was often there. You just stopped seeing it.

• • •

TAKE ACTION: YOUR NEXT STEPS

Reflection

Think about your current work. Answer these questions honestly:

1. What contributions have you made recently that you discounted? List three things you accomplished this week that you dismissed as "not counting." Be specific.

2. Whose problems does your work solve? Name actual people—colleagues, clients, users, customers—who benefit from what you do. Make it concrete.

3. What's driving your work—avoidance or values? Are you primarily running from negative consequences or running toward something that matters? Be honest.

4. What recognition are you waiting for? What validation do you need to feel your work matters? What would happen if that recognition never came?

5. What capacity is your current role building? What skills, experiences, or capabilities are you developing—even if the role isn't where you want to end up?

Implementation

Keep a contribution log for one week. Each day, write down three things you contributed. They don't have to be big. Help you gave. Problems you solved. Things you made better. At the end of the week, review the log. Notice how much you actually contributed that you would otherwise have forgotten.

Connect tasks to people. Before starting your next significant task, ask: Who will benefit from this? Name them if possible. Complete the task with that person in mind. Notice if the experience of the work changes.

Practice crediting yourself. At the end of each workday this week, complete this sentence: "Today I contributed by..." Don't discount. Don't minimize. Just name what you did.

Troubleshooting

"My work genuinely doesn't matter. I'm a cog in a machine producing nothing of value."

This might be true. Some work really is disconnected from meaningful contribution. If that's the case, the question becomes: What are you going to do about it? The shifts in this chapter help you find meaning where it exists. They can't create meaning where none is possible. If your honest assessment is that your current work serves nothing you value, that's important information about what needs to change.

"I do recognize my contributions, but it doesn't help the emptiness."

The emptiness might be pointing to something deeper than cognitive framing. You might be in the wrong role, the wrong field, or dealing with burnout that requires more than perspective shifts. The shifts in this chapter are tools, not solutions to every form of work dissatisfaction. If you've genuinely reframed your work and still feel empty, that signal deserves attention.

"I can't afford to care less about recognition. My career advancement depends on being visible."

True—career advancement often requires visibility and recognition. The shift isn't about abandoning that strategic reality. It's about where you locate your sense of meaning. You can pursue recognition strategically while not depending on it for your sense of worth. Do the work because it's worth doing. Also, make sure the right people see it. Both/and.

• • •

Meaning in work is partly about the work itself and partly about how you see the work. Some roles genuinely offer more inherent meaning than others. But within any role, there's a wide range of possible experience—from empty drudgery to engaged contribution—depending on the frame you hold.

The shifts in this chapter don't promise that all work will feel meaningful. They offer a way to stop filtering out the meaning that's already there.

With Part Two complete, we've covered the career domain: failure, criticism, uncertainty, comparison, and meaning. These are the cognitive battlegrounds where professional lives are won and lost—not through talent alone, but through the interpretations you bring to inevitable challenges.

Part Three turns to relationships. The cognitive patterns that create suffering in your career also show up in your connections with others—but the specific triggers are different. Conflict, disappointment, other people's behavior, your own patterns, and the challenge of intimacy all present unique opportunities for distorted thinking and unique opportunities for reframing. That's where we're headed next.

PART THREE

RELATIONSHIP SHIFTS

CHAPTER 8: Reframing Conflict and Disagreement

The cognitive shifts that transform friction into connection

> *When you wake up in the morning, tell yourself: The people I deal with today will be meddling, ungrateful, arrogant, dishonest, jealous, and surly. They are like this because they can't tell good from evil. But I have seen the beauty of good, and the ugliness of evil, and have recognized that the wrongdoer has a nature related to my own.*
>
> — Marcus Aurelius, Meditations

You're in the middle of an argument with your partner about something that shouldn't even be a big deal. Maybe it's the dishes. Maybe it's whose family to visit for the holidays. Maybe it's how to handle a situation with a mutual friend.

Somewhere in the exchange, you stopped listening. You started preparing your rebuttal. They said something that triggered you, and now you're not trying to understand—you're trying to win. The original issue is almost forgotten. This has become about who's right.

An hour later, the argument sputters out. Nothing is resolved. You're both hurt and defensive. The connection that was there before the argument is now strained. And you're left wondering: How did we get here? Why does this keep happening?

The answer is in how you framed the conflict.

Most relationship friction isn't caused by the actual issue being discussed. It's caused by what each person's mind does with the issue—the interpretations, attributions, and cognitive patterns that transform disagreement into warfare.

This chapter is about a different way through conflict. Not avoiding it—that creates its own problems. But navigating it without losing the connection you're trying to protect.

• • •

Black-and-White Thinking in Relationships

When conflict triggers your threat response, your thinking narrows. Nuance disappears. The other person becomes entirely wrong, and you become entirely right.

This is black-and-white thinking—the collapse of complex situations into binary categories. Right or wrong. Good or bad. With me or against me.

In a disagreement, black-and-white thinking sounds like this:

"You always do this."

"You never listen."

"This is completely your fault."

"I'm not wrong at all."

These statements feel true in the moment. They're not. They're the emotional brain painting in absolutes because absolutes are easier to process than complexity.

Your partner doesn't always do that thing. There have been times they didn't. You're not entirely blameless. There were moments you contributed to the escalation. The situation isn't completely one person's fault. It's a dance, with moves from both sides.

Black-and-white thinking is dangerous in relationships because it forecloses understanding. If you're entirely right and they're entirely wrong, there's nothing to learn and no room for movement. You're not trying to solve a problem together—you're prosecuting a case. And when both parties are prosecuting cases, no one's listening.

The alternative isn't abandoning your position. It's holding your position while acknowledging that their position also exists, also makes some kind of sense from their perspective, and isn't evidence of malice or stupidity.

• • •

The Fundamental Attribution Error

When your partner does something that bothers you, your mind generates an explanation. And that explanation is almost certainly distorted.

The fundamental attribution error is one of the most robust findings in social psychology. It works like this: when others behave badly, we attribute it to their character. When we behave badly, we attribute it to our circumstances.

Your partner snaps at you. You think: They're being hostile. They're inconsiderate. They have an anger problem.

You snap at your partner. You think: I was stressed. I'd had a terrible day. The situation triggered me.

Same behavior. Completely different explanations. And the asymmetry creates endless conflict.

When you attribute their behavior to character, you're not seeing a person in a moment—you're seeing a fixed trait. They're not acting impatient; they are impatient. They're not feeling frustrated; they are a frustrating person. This framing makes the behavior seem permanent and intentional. It invites resentment rather than compassion.

When you attribute your own behavior to circumstances, you give yourself context. You understand why you acted that way. You see it as situational, temporary, not representative of who you really are.

The correction is to extend to others the contextual understanding you naturally give yourself. When they snap at you, ask: What circumstances might explain this? Are they tired? Stressed? Dealing with something I don't know about? The behavior is the same. The interpretation changes everything.

This doesn't mean excusing harmful behavior. Some actions need to be addressed regardless of circumstances. But the fundamental attribution error isn't about excusing—it's about understanding. And understanding is the prerequisite for productive conflict.

• • •

Perspective-Taking Reframing

Most conflicts escalate because both parties are arguing from their own perspective while failing to understand the other's.

You see the situation from where you stand. From there, your position is obviously correct. The evidence supports you. Your reasoning is sound. Why can't they see it?

They see the situation from where they stand. From there, their position is obviously correct. The evidence supports them. Their reasoning is sound. Why can't you see it?

Both views are incomplete. Neither person has access to the full picture. The conflict isn't really about who's right—it's about two people with different information, different priorities, and different frameworks arriving at different conclusions.

Perspective-taking reframing involves deliberately imagining the situation from their point of view. Not as an exercise in capitulation—you don't have to agree. But as an exercise in comprehension. What do they see that you might be missing? What matters to them that you might be discounting? What experiences have they had that make their view make sense to them?

This is harder than it sounds. In the heat of conflict, your brain doesn't want to take their perspective. It wants to fortify your position. Perspective-taking feels like weakness, like giving ground.

But it's not. It's intelligence-gathering. You can't respond effectively to something you don't understand. And you can't understand their position if you refuse to inhabit it even momentarily.

The practice works like this: Before responding in conflict, pause and ask, "How does this look from their side?" Answer the question seriously. Then respond—not from a place of pure self-interest, but from a place of understanding both perspectives.

• • •

Epictetus on Responding to Difficult People

The Stoics assumed difficult people were inevitable. Not as an occasional annoyance but as a constant feature of life. Their advice wasn't to avoid conflict but to manage your response to it.

Epictetus offered this perspective: when someone acts badly toward you, remember that they're acting according to their view of what's right. They're not trying to be wrong—they're doing what seems correct to them given their understanding.

"When someone acts unjustly toward you, consider what view of good or evil led them to act that way. Once you see this, you'll feel compassion rather than surprise or anger. For either you yourself still share their view of what's good, or you've come to see something better. If the former, you should forgive them. If the latter, you should pity them."

This isn't about excusing bad behavior. It's about understanding its source. People don't typically wake up intending to be difficult. They act based on their beliefs, priorities, and interpretations—which may be flawed, but are their best understanding.

The Stoic response to difficult people involves three elements:

First, **expectation management**. Marcus Aurelius began each day expecting to encounter difficult people. This wasn't pessimism—it was preparation. When difficulty arrives, you're not surprised or outraged. You're ready.

Second, **understanding their view**. What perspective makes their behavior make sense to them? You don't have to share that perspective, but you can understand it. This transforms the person from an adversary into a human operating with different information.

Third, **controlling your response**. You can't control what they do. You can control what you do. Your response is your domain—the only thing in the interaction that's truly up to you.

• • •

The Four Conflict Shifts

SHIFT 27

> **FROM:** *"They're wrong."*
>
> **TO:** *"They see it differently."*

This shift attacks black-and-white thinking at its source.

"They're wrong" is a conclusion disguised as an observation. It assumes that your perspective is correct, and any deviation from it is error. But in most relationship conflicts, both parties have access to partial information. Neither view is complete. Neither is simply "right."

"They see it differently" acknowledges disagreement without dismissing their view as wrong. They're looking at the same situation from a different vantage point. They have different information, different priorities, different interpretations. The difference doesn't make them wrong—it makes them different.

This shift doesn't require abandoning your position. You can believe you're correct while acknowledging that their contrary view makes sense from where they stand. Both things can be true: you see it one way, they see it another, and both of you have reasons for your views.

The practical effect is reduced righteousness. When you're certain they're wrong, you stop listening and start lecturing. When you recognize they see it differently, you stay curious. What are they seeing that you're missing? What's their experience of this situation? The questions open space for resolution.

SHIFT 28

> **FROM:** *"We need to agree."*
>
> **TO:** *"We need to understand."*

This shift relocates the goal of conflict.

"We need to agree" makes resolution contingent on convergence. One of us has to change positions. Either you come to my view or I come to yours. Until that happens, the conflict continues.

But many disagreements don't require agreement. They require understanding. You can hold different views while respecting each other's positions. You can disagree without the disagreement poisoning the relationship.

"We need to understand" makes the goal comprehension rather than capitulation. I need to understand why you see it this way. You need to understand why I see it this way. Once we understand each other, we can figure out how to move forward—even if we never fully agree.

This matters because agreement isn't always possible. Some differences are genuine. You have different values, different priorities, different tolerances. Forcing agreement means someone has to betray their actual view. Understanding allows both views to exist while finding workable paths forward.

The conversation shifts from "Here's why you should agree with me" to "Help me understand how you see this." The tone changes. The defensiveness decreases. And often, paradoxically, understanding makes agreement more likely—because people soften when they feel heard.

SHIFT 29

FROM: *"This argument proves something is broken."*

TO: *"This disagreement proves we both care."*

This shift reframes conflict itself.

"This argument proves something is broken" treats conflict as pathology. Happy couples don't fight. Strong relationships don't have disagreements. If we're arguing, something must be wrong with us.

But conflict isn't evidence of brokenness. It's evidence of engagement. Indifferent people don't argue—they disengage. The presence of conflict means both parties are invested enough to struggle.

"This disagreement proves we both care" recognizes the energy in conflict. You're both fighting because the issue matters. You're both engaged because the relationship matters. The heat comes from caring, not from dysfunction.

This reframe doesn't excuse destructive conflict patterns. Some couples argue in ways that genuinely are harmful—with contempt, cruelty, or abuse. Those patterns need to be addressed. But the existence of disagreement itself isn't the problem. How you handle disagreement is the variable.

When you see conflict as proof of caring, you enter it differently. The argument isn't evidence that your relationship is failing. It's a moment where two people who care about each other are struggling to reconcile different views. That struggle, navigated well, can strengthen the connection rather than eroding it.

SHIFT 30

> **FROM:** *"I need to win this."*
>
> **TO:** *"I need to stay connected through this."*

This shift redirects the objective of conflict.

"I need to win this" turns your partner into an opponent. The conflict becomes a competition with a winner and a loser. Your job is to defeat their arguments, establish your position, and emerge victorious.

But what happens when you win an argument with someone you love? They lose. And their loss doesn't feel like your victory—it damages the relationship. The connection suffers. One person wins the argument and both people lose the intimacy.

"I need to stay connected through this" makes the relationship the priority instead of the position. Yes, you disagree. Yes, you have a perspective you believe in. But the goal isn't to crush their position—it's to navigate the disagreement without sacrificing the connection.

This changes how you argue. You can still advocate for your view. You can still explain your reasoning. But you're not trying to destroy them. You're trying to stay in relationship while working through difference.

Practically, this means checking your tone. It means avoiding contempt, eye-rolling, and dismissiveness. It means continuing to treat them as someone you love even when you're frustrated with them. The disagreement is temporary. The relationship is what you're trying to protect.

• • •

Putting It Together

When conflict arises, you have a sequence available.

Notice the escalation. Feel the defensiveness rising. Notice when you stop listening and start preparing rebuttals. Recognize when the original issue gets lost in the competition.

Check for distortions. Are you in black-and-white thinking? Are you making the fundamental attribution error—attributing their behavior to character while excusing your own with circumstances? Are you assuming malice where there might be misunderstanding?

Apply perspective-taking. Pause and genuinely ask: How does this look from their side? What information do they have that I might be missing? What makes their view make sense to them?

Choose the shifts. They see it differently—that's not wrong, just different. We need to understand, not necessarily agree. This disagreement proves we care. The goal is to stay connected through this, not to win at the cost of connection.

Engage differently. With the reframes in place, re-enter the conversation. Listen more. Defend less. Seek understanding before seeking victory.

This won't eliminate conflict. Disagreement is part of any real relationship. But it changes the nature of conflict—from warfare that damages connection to friction that, handled well, can actually strengthen it.

• • •

TAKE ACTION: YOUR NEXT STEPS

Reflection

Think of a recent conflict with someone close to you—a partner, family member, or close friend. Work through these questions:

1. What was your position? Summarize your view of the situation. What did you think was right?

2. What was their position? Summarize their view—not in a dismissive way, but as they would describe it themselves. Can you articulate it fairly?

3. Where did black-and-white thinking show up? Did you use words like "always," "never," "completely," or "entirely"? Did you reduce the situation to binary categories?

4. What circumstances might explain their behavior? Apply perspective-taking. What might they have been dealing with that you didn't consider? What makes their position make sense from where they stand?

5. What would staying connected have looked like? If your goal had been connection rather than winning, how might you have engaged differently?

Implementation

Practice the pause-and-ask technique. In your next disagreement, when you feel the urge to respond defensively, pause. Silently ask: "How does this look from their side?" Then respond. The pause creates space for the shift.

Replace "You're wrong" with "Help me understand." This week, whenever you feel the impulse to tell someone they're wrong, replace it with a request for understanding. "Help me understand why you see it that way." Notice how the conversation changes.

Name the goal out loud. In your next significant disagreement, say something like: "I want to make sure we stay connected through this, even if we don't agree." Making the goal explicit changes the dynamic for both parties.

Troubleshooting

"But sometimes they really are wrong. Shouldn't I be able to say that?"

You can believe they're wrong while still recognizing that they see it differently. The shift isn't about abandoning your view—it's about how you hold your view. You can think you're right without dismissing their perspective as simply error. And practically, "You're wrong" rarely persuades anyone. "Help me understand" opens dialogue.

"What if they're not interested in understanding—they just want to win?"

You can't control their approach, only yours. If you enter with the understanding frame, you change the dynamic—even if they don't reciprocate initially. Sometimes your shifted approach invites them to shift too. Sometimes it doesn't. But staying connected is still worth attempting from your side.

"Some conflicts really are zero-sum. One of us has to give in."

True—some decisions require choosing one option over another. But even in zero-sum situations, how you navigate the conflict matters. You can "win" in a way that damages the relationship or "win" in a way that preserves it. The shifts help you pursue the latter.

• • •

Conflict is inevitable in any close relationship. The question isn't whether you'll disagree, but how you'll handle disagreement when it comes.

The shifts in this chapter offer a different relationship with conflict. Not avoidance—that creates distance. But a way through friction that maintains connection. Understanding rather than defeating. Staying close rather than winning.

In the next chapter, we turn to expectations and disappointment. Before conflict even erupts, something else is happening: you expected something from someone, and they didn't deliver. That gap between expectation and reality is where much relationship suffering lives. The cognitive skills for navigating it are specific and learnable.

CHAPTER 9: Reframing Expectations and Disappointment

How to want things without being destroyed when you don't get them

> *Begin each day by telling yourself: Today I shall meet with interference, ingratitude, insolence, disloyalty, ill-will, and selfishness—all of them due to the offenders' ignorance of what is good or evil.*
>
> — Marcus Aurelius, Meditations

Your birthday is coming up. You haven't said anything, but you expect your partner to plan something. Not a party—just acknowledgment. Dinner out. A thoughtful gift. Some indication that the day matters to them because you matter to them.

The day arrives. They got you a card—nice, but generic. They suggest ordering pizza and watching a movie. They seem pleased with themselves. They have no idea that you're crushed.

You don't say anything. But the wound festers. They should have known. You've been together for years. How could they not know this mattered? How could they be so thoughtless?

The resentment builds. You're cold for days afterward. They don't understand why. You don't explain because you believe you shouldn't have to. If they loved you, they would have known.

This is the anatomy of disappointment. And almost every piece of it is constructed from cognitive distortions.

• • •

Unspoken Contracts and the Resentment They Breed

Most relationship disappointment comes not from what people did, but from what they failed to do. And most of that failure involves expectations they never knew about.

Unspoken contracts are agreements that exist only in your mind. You've decided what the other person should do, but you haven't told them. You've established standards for their behavior that they haven't agreed to. Then, when they fail to meet standards they didn't know existed, you feel betrayed.

The birthday scenario is classic. You expected something. You didn't say what. They delivered something else. Now you're hurt—and they don't even know why.

This pattern shows up everywhere in relationships:

"They should have offered to help."

"They should have asked how my meeting went."

"They should have remembered I don't like that."

"They should have known I was struggling."

Each "should" represents an expectation the other person may not share or even be aware of. Your disappointment is real. But the contract they violated was never signed—it was imagined.

The resentment that follows is particularly corrosive because it feels justified. Of course you're upset. They should have done that thing. The fact that they didn't know about your expectation feels irrelevant. If they cared enough, they would have figured it out.

But people can't read minds. Even people who love you deeply will fail to meet expectations they don't know about. Your unspoken contract is invisible to them. And holding them accountable for invisible agreements is a recipe for chronic disappointment.

• • •

The Fairness Fallacy

Beneath many unspoken contracts lies a deeper distortion: the belief that relationships should be fair.

The fairness fallacy assumes that effort, sacrifice, and investment should be reciprocated equally. If I do something for you, you should do something equivalent for me. If I give, you should give back. The relationship should be balanced, like a ledger.

This framing creates constant scorekeeping. You track what you've contributed and compare it to what they've contributed. Any imbalance feels like injustice. They're not pulling their weight. You're doing more than your share. It's not fair.

The problem is that relationships aren't accounting exercises. People have different capacities, different circumstances, different ways of expressing care. What feels like equivalent effort to one person might look completely different to another.

Your partner might show love through acts of service while you show it through words of affirmation. You're both contributing, but in currencies the other doesn't fully value. The ledger never balances because you're measuring in different units.

Beyond the measurement problem, the fairness framing itself is flawed. Relationships aren't transactions. You don't love someone because you expect equal return. You love them because you choose to. Making that choice contingent on balanced reciprocation transforms love into commerce.

This doesn't mean accepting genuinely exploitative relationships. If you're consistently giving while receiving nothing—if the imbalance is chronic and draining—that's information about the relationship's viability. But the fairness fallacy operates even in healthy relationships, creating resentment where none is warranted.

• • •

The Stoic Practice of Examining Impressions

The Stoics recognized that between experience and reaction lies a gap—a moment where interpretation happens. They called this "examining impressions."

When someone disappoints you, the disappointment feels immediate. They did something; you felt hurt. Cause and effect. But the Stoics argued that the disappointment isn't automatic. It follows from judgments you made about what happened.

Your partner gave you a generic birthday. That's the event. The hurt comes from the interpretation: They don't care enough. They should have known better. They failed a test they didn't know they were taking.

Examining impressions means pausing before the reaction and questioning the interpretation. Is it true that they don't care? Or did they care in a way that doesn't match your expectations? Is it true they should have known? Or is that an assumption you made?

Epictetus put it bluntly: "It is not things that disturb us, but our judgments about things." Your partner's gift didn't hurt you. Your judgment about what the gift meant hurt you.

This isn't about denying feelings or suppressing hurt. It's about accuracy. If your interpretation is accurate, the hurt is proportionate. If your interpretation is distorted, you're suffering more than the situation warrants.

The practice involves three steps:

First, **notice the impression**. Something happened; you're having a reaction. Name the interpretation your mind generated. "I'm interpreting this as evidence that they don't care."

Second, **examine the impression**. Is this interpretation accurate? Is it the only possible interpretation? What else could explain what happened?

Third, **choose your response**. Based on a more accurate interpretation, what response makes sense? This might still involve hurt—but informed hurt, proportionate to what actually happened.

• • •

ACT Acceptance in Intimate Relationships

Acceptance and Commitment Therapy offers a framework particularly useful for relationship disappointment: acceptance.

Acceptance in ACT isn't resignation or approval. It's willingness to experience reality as it is rather than fighting against what is. You can accept something while also working to change it. But you can't change anything you refuse to see clearly.

Applied to relationships, acceptance means accepting that other people are who they are—not who you want them to be, not who you expected them to be, but who they actually are.

Your partner has a certain capacity for emotional expression. You might wish it were different. You might prefer someone more effusive, more attentive, more demonstrative. But wishing doesn't change their nature. The person in front of you is the person they are.

Acceptance involves recognizing the gap between your expectations and their reality—and letting go of the demand that they bridge that gap. Not because your preferences don't matter, but because demanding that someone be different than they are is a losing battle.

This might mean adjusting your expectations to match their capacity. It might mean communicating your needs more clearly so they know what you want. It might mean accepting that certain needs won't be met in this relationship and finding other sources to meet them. In some cases, it might mean recognizing that the gap is too large and the relationship isn't right for you.

All of these responses start with acceptance. You can't navigate reality effectively if you're arguing with it.

• • •

The Four Disappointment Shifts

SHIFT 31

FROM: *"They should have known."*

TO: *"I could have communicated."*

This shift addresses the mind-reading expectation.

"They should have known" assumes that caring automatically produces understanding. If they loved you, they would intuit your needs. If they paid attention, they would figure out your expectations. The need to explain feels like failure—theirs, for not knowing, and somehow yours, for having to tell them.

But people can't read minds. Even devoted partners miss signals. Even attentive friends overlook needs. The expectation that they should know without being told sets them up to fail and you up to be disappointed.

"I could have communicated" relocates agency. You have needs and expectations. The most reliable way for others to know about them is for you to say what they are. Not hinting. Not expecting them to figure it out. Directly stating what you want.

This doesn't mean every need must be spelled out in advance. Some things genuinely are obvious, and failure to notice them might indicate real inattention. But in most cases of "they should have known," honest reflection reveals that you never actually told them.

The shift invites a different response to disappointment. Before resentment, ask: Did I communicate this expectation clearly? If not, the disappointment might be an invitation to communicate—not evidence of their failure to care.

SHIFT 32

FROM: *"They let me down."*

TO: *"My expectations weren't aligned with their capacity."*

This shift reframes disappointment as misalignment rather than betrayal.

"They let me down" treats the unmet expectation as a personal failure on their part. They had a job to do and didn't do it. They dropped the ball. The framing implies willful neglect—they could have met your expectation and chose not to.

But often, they couldn't. Not because they didn't care, but because your expectation exceeded their capacity. They're not as emotionally expressive as you need them to be.

They're not as attentive to details as you expected. They're not as reliable as you assumed.

"My expectations weren't aligned with their capacity" separates expectation from delivery without assigning blame. You expected something. They didn't deliver it. The gap might be because they failed—but it might also be because your expectation didn't match what they're capable of providing.

This shift invites examination of both sides. Was your expectation reasonable? Was it communicated? Did it account for who they actually are, or who you wanted them to be? The disappointment is real, but the cause might be misalignment rather than betrayal.

From this framing, you have options. You can adjust your expectations. You can communicate more clearly. You can decide whether the gap is acceptable or not. All of these are more productive than nursing resentment over being "let down."

SHIFT 33

FROM: *"This isn't fair."*

TO: *"Fairness isn't guaranteed; my response is my choice."*

This shift confronts the fairness fallacy directly.

"This isn't fair" protests against an imbalance. You gave more than you received. You deserved better than you got. The ledger doesn't balance, and that's unjust.

The protest is understandable. Fairness is a deep human value. We're wired to notice and resent imbalances, especially when we're on the short end.

But relationships don't promise fairness. People have different capacities for giving. Circumstances change what each person can contribute. Some periods are imbalanced in one direction; others shift. The expectation of constant fairness sets you up for constant disappointment.

"Fairness isn't guaranteed" accepts this reality. It doesn't mean tolerating exploitation. It means releasing the demand that every exchange balance perfectly. Some imbalances are temporary. Some reflect different giving styles. Some are simply the nature of relationships between different people.

"My response is my choice" returns agency. Whether the situation is fair or not, you choose what to do about it. You can communicate your needs. You can adjust your giving. You can leave if the imbalance is chronic and damaging. What you can't do—productively—is demand that the universe deliver fairness.

SHIFT 34

FROM: *"They owe me."*

TO: *"I gave freely; what they do now is their choice."*

This shift dismantles covert contracts.

"They owe me" implies that your giving created a debt. You did something for them; now they're obligated to reciprocate. Your generosity wasn't free—it was an investment that expects return.

But did they know that? Did they agree to the terms? Or did you unilaterally decide that your gift created their obligation?

Most "they owe me" resentment comes from covert contracts—expectations of reciprocation that were never made explicit. You gave with strings attached, but you didn't tell them about the strings. Now you're angry that they haven't paid a bill they didn't know they owed.

"I gave freely" reclaims the gift as a gift. If you gave something, it was your choice to give. They didn't force you. If you gave expecting something in return, that expectation was yours—possibly unspoken, possibly unknown to them.

"What they do now is their choice" releases the demand for reciprocation. They might give back. They might not. That's their decision, not their debt. If reciprocation is important to you, you can discuss it explicitly. But holding them accountable for a contract they never agreed to isn't fair—even if you feel you've given more.

This doesn't mean accepting one-sided relationships. If someone consistently takes without giving, that's information about who they are. But the solution isn't to pile up unspoken debts and resent them for not paying. The solution is to stop giving to people who don't reciprocate—or to give truly freely, without expectation.

• • •

Putting It Together

When disappointment strikes, you have a sequence available.

Notice the expectation. What did you expect that didn't happen? Be specific. Name the unmet expectation clearly.

Examine whether it was communicated. Did they know about this expectation? Did you tell them directly, or did you assume they would figure it out?

Check for distortions. Are you assuming mind-reading? Applying the fairness fallacy? Holding them to a contract they never signed?

Apply the relevant shift. I could have communicated. My expectations weren't aligned with their capacity. Fairness isn't guaranteed. I gave freely.

Choose your response. Based on a clearer view, what makes sense now? Communication? Adjustment of expectations? Acceptance? A conversation about what you need going forward?

This won't eliminate disappointment. People will continue to fall short of expectations, sometimes yours and sometimes their own. But you can reduce unnecessary suffering—the suffering that comes from distorted interpretations rather than genuine failures. And you can respond to real disappointments more effectively, without the fog of resentment obscuring your options.

• • •

TAKE ACTION: YOUR NEXT STEPS

Reflection

Think of a recent disappointment with someone close to you—a time they didn't meet your expectations. Work through these questions:

1. What was the expectation? State it specifically. What did you expect them to do or not do?

2. Was the expectation communicated? Did you tell them directly what you wanted? Or did you assume they would know?

3. Was the expectation reasonable given their capacity? Does this person typically deliver what you expected? Or does the expectation exceed what they've demonstrated they can provide?

4. What covert contract might be operating? Did you give something expecting reciprocation? Did they know about that expectation?

5. What would acceptance look like here? If you accepted this person as they are—not as you wish they were—how would that change your response to the disappointment?

Implementation

Practice explicit communication. This week, identify one expectation you have of someone close to you that you haven't stated directly. State it. "I would really appreciate it if you would..." Notice how clear communication changes the dynamic.

Audit your covert contracts. Review your recent resentments. For each one, ask: Did they know about this expectation? Did they agree to it? If the contract was never made explicit, consider releasing the resentment—or making the contract explicit going forward.

Give one thing with no strings attached. This week, do something generous for someone with genuinely zero expectation of return. Notice how it feels to give freely, without the weight of expected reciprocation.

Troubleshooting

"But some things really should be obvious. I shouldn't have to tell them basic things."

Maybe. But "obvious" is relative. What's obvious to you might not be obvious to them, given their background, experiences, and mental model of relationships. The question isn't whether you should have to tell them—it's whether telling them would solve the problem. If yes, why not tell them? Principle is cold comfort when communication would actually help.

"What if they keep disappointing me even when I communicate clearly?"

That's important data. If you've communicated clearly, multiple times, and they still don't deliver—that tells you something about their capacity or priorities. At that point, the question shifts from "How do I get them to meet my expectations?" to "Can I accept this gap, or is it a dealbreaker?" Both are valid responses.

"Isn't it unfair to just accept unfairness? Shouldn't I advocate for myself?"

Absolutely advocate for yourself. The shift isn't about accepting mistreatment. It's about releasing the demand that the universe deliver perfect fairness. You can advocate, communicate, set boundaries, and even leave relationships that don't serve you—without the additional suffering of protesting against reality. Acceptance isn't passivity. It's clarity about what is, which enables more effective action.

• • •

Expectations are unavoidable. You can't care about people without wanting things from them and for them. But unexamined expectations—especially unspoken ones—create a minefield of potential disappointment.

The shifts in this chapter help you navigate that minefield. Not by wanting less, but by holding expectations more lightly. Communicating more clearly. Examining whether your disappointments are proportionate to what actually happened. Releasing the demand that others read your mind or that life be perfectly fair.

In the next chapter, we turn to other people's behavior—not just when they disappoint expectations, but when they act in ways that frustrate, confuse, or hurt you. The cognitive patterns that generate suffering around others' behavior are specific and identifiable. So are the shifts that liberate you from being controlled by what other people do.

CHAPTER 10: Reframing Other People's Behavior

The liberating shift from trying to control others to managing your own responses

> *How much more grievous are the consequences of anger than the causes of it.*
>
> — Marcus Aurelius, Meditations

Your mother criticizes how you're raising your kids. Again. The same comments, the same tone, the same implication that you're doing it wrong. You've asked her to stop. You've explained why it bothers you. You've had the conversation multiple times. Nothing changes.

Why does she keep doing this? Doesn't she see how it affects you? Doesn't she care about your relationship? If she would just stop, everything would be fine.

You've spent years trying to get her to change. You've reasoned with her. You've gotten angry. You've withdrawn. You've tried guilt, boundaries, distance. Sometimes there's temporary improvement. Then the behavior returns, and so does your frustration.

Here's the uncomfortable truth: you cannot control what she does. You have never been able to. And every moment you spend trying is a moment spent on something impossible—while ignoring the one thing you actually can change.

• • •

The Control Fallacy in Relationships

One of the most persistent sources of relationship suffering is the belief that you can control other people.

You can't. Not really. You can influence, persuade, incentivize, request. You can set boundaries and enforce consequences. But the actual decision to change—that belongs to them. And many people choose not to, regardless of what you do.

The control fallacy manifests as the belief that if you just find the right approach, say the right words, apply the right pressure, they'll change. The problem is framed as tactical: you haven't found the key yet. Once you do, the door will open.

But some doors don't have keys. Some people won't change no matter what you do. Not because you failed to find the right approach—because their choices are theirs, and they've made them.

This is hard to accept because it feels like giving up. If you stop trying to change them, doesn't that mean accepting behavior that hurts you? Doesn't that mean letting them win?

But accepting that you can't control them isn't the same as accepting their behavior. You still have options. You can change your own behavior. You can limit contact. You can leave the relationship. You can change how you respond to their actions. All of these are things you actually control.

What you can't do—what has never been possible—is reach inside their head and flip the switch that makes them different. That switch is theirs to flip. And they might never flip it.

• • •

Codependence and Cognitive Distortions

The compulsion to control others often has deeper roots than simple frustration. Codependence is a pattern in which your sense of wellbeing becomes excessively dependent on someone else's behavior.

In codependent dynamics, you don't just want them to change—you need them to change for you to be okay. Your emotional state is hostage to their choices. When they behave well, you feel good. When they don't, you fall apart.

This creates a desperate investment in controlling their behavior. You can't just let them be who they are because who they are determines whether you're happy. The stakes feel existential.

Several cognitive distortions fuel codependence:

Personalization: You interpret their behavior as being about you. They drink because you're not enough. They're angry because you did something wrong. Their problems become your responsibility.

Mind-reading: You assume you know what they need better than they do. If they would just listen to you, their life would improve. Your vision of their wellbeing becomes more important than their own choices.

Emotional reasoning: You feel responsible for their happiness, so you assume you are responsible. The feeling of obligation is taken as evidence of actual obligation.

Control fallacy: You believe that with enough effort, you can fix them. Their failure to change is your failure to find the right intervention.

Breaking free from codependence involves recognizing these distortions and reclaiming appropriate boundaries. Their behavior is theirs. Their choices are theirs. Your job is managing your own life—not managing theirs.

• • •

The Reattribution Technique

When someone does something that bothers you, your mind instantly generates an explanation. And that explanation usually focuses on you.

They're doing this because they don't respect me. They're doing this to hurt me. They're doing this because they don't care what I think.

The reattribution technique involves deliberately generating alternative explanations—ones that don't center on you.

Your mother criticizes your parenting. Your automatic interpretation: She thinks I'm a bad parent. She doesn't respect my choices. She's trying to undermine me.

Reattribution asks: What other explanations might fit the facts?

Maybe she's anxious about her grandchildren and doesn't know how to express it constructively. Maybe she's repeating patterns from how she was raised and lacks self-awareness about it. Maybe she's struggling with feeling irrelevant as she ages and criticism is how she maintains involvement. Maybe she genuinely believes she's being helpful and doesn't realize how it lands.

None of these explanations excuse the behavior. You can still wish she would stop. But they change your relationship to the behavior. You're not the target of deliberate attack—you're dealing with a person acting out their own struggles.

This matters because the explanation you hold determines your emotional response. If she's attacking you, you feel defensive and angry. If she's struggling with her own issues in a way that happens to affect you, you might feel frustrated but less personally wounded.

Reattribution doesn't require knowing which explanation is correct. The point is recognizing that your automatic explanation isn't the only possibility—and might not be the most accurate.

• • •

Marcus Aurelius on Expecting Difficult Behavior

Marcus Aurelius was emperor of Rome. He dealt with difficult people constantly—senators with agendas, generals with egos, advisors with conflicting interests, family members with complicated motives. His response wasn't to control them. It was to prepare himself.

Each morning, he would remind himself that he would encounter frustrating people:

"Begin each day by telling yourself: Today I shall meet with interference, ingratitude, insolence, disloyalty, ill-will, and selfishness—all of them due to the offenders' ignorance of what is good or evil."

This wasn't pessimism. It was psychological preparation. By expecting difficult behavior, he wasn't surprised when it appeared. By attributing it to ignorance rather than malice, he maintained compassion instead of rage.

Marcus also recognized the futility of trying to change others:

"Don't demand that things happen as you wish, but wish that they happen as they do happen, and you will go on well."

He applied this to people as much as to events. You can wish people would behave differently. You can even work to influence them. But demanding that they be different than they are is a recipe for suffering.

The Stoic approach to difficult people involves three elements:

Expectation: Know in advance that people will be difficult. This is not cynicism—it's realism. Humans are complicated. They act on confused motives. They repeat patterns. Expecting difficulty means you're not blindsided by it.

Understanding: Try to understand why they do what they do. Not to excuse it, but to see it clearly. Their behavior makes sense to them, given their beliefs and circumstances. Understanding that doesn't require approving.

Control focus: Direct your energy to what you can actually affect—your own responses. Their behavior is their domain. Your response is yours. The only productive question is: How will I handle this?

• • •

The Four Other-People Shifts

SHIFT 35

FROM: *"Why do they keep doing that?"*

TO: *"That's what they do; how will I respond?"*

This shift moves from frustrated bewilderment to practical adaptation.

"Why do they keep doing that?" is a rhetorical question that spirals into rumination. You're not genuinely seeking understanding—you're expressing frustration. The question implies that their behavior is incomprehensible, that it shouldn't be happening, that you're baffled by their persistence.

But you're not really baffled. You know exactly what they do. You've seen the pattern dozens of times. The surprise isn't genuine—it's performed outrage at a predictable reality.

"That's what they do" accepts the pattern as data. This person, in this situation, behaves this way. That's established. Maybe it's unfortunate. Maybe you wish it were different. But the pattern is real.

"How will I respond?" redirects to your domain. You can't change what they do—not directly. You can change what you do. How will you handle this behavior? What boundaries will you set? What response serves your interests and values?

The shift transforms helpless frustration into active problem-solving. You're no longer banging your head against their immovability. You're adapting to a reality that, however frustrating, you can navigate.

SHIFT 36

FROM: *"They're doing this to hurt me."*

TO: *"They're doing this for their reasons."*

This shift applies reattribution to hurtful behavior.

"They're doing this to hurt me" assumes intentional malice. Their behavior is targeted. You are the point. They wake up thinking about how to cause you pain.

Sometimes this is true. Some people are deliberately cruel. But more often, it's not. More often, their behavior has little to do with you. They're acting on their own fears, insecurities, habits, and limitations. You happen to be affected, but you're not the purpose.

"They're doing this for their reasons" separates the impact from the intent. Their behavior hurts you—that impact is real. But the intent might be entirely different from causing hurt. They might be seeking control, acting out anxiety, repeating childhood patterns, or simply being thoughtless.

This matters because intent changes the meaning of the behavior. Deliberate cruelty is a betrayal. Thoughtlessness is frustrating but different. Self-protective behavior that happens to hurt you is understandable even when painful.

The shift doesn't excuse the behavior or minimize your hurt. It clarifies what you're actually dealing with. And clarity is the prerequisite for effective response.

SHIFT 37

FROM: *"If they loved me, they would change."*

TO: *"Love doesn't guarantee change."*

This shift addresses one of the most painful beliefs in relationships.

"If they loved me, they would change" makes love and behavior change equivalent. They do the thing that hurts you. If they loved you enough, they would stop. The fact that they haven't stopped means they don't love you enough.

The logic is clean but wrong.

Love doesn't automatically produce behavioral change. People can love deeply and still repeat patterns that hurt the people they love. Addiction, anxiety, depression, trauma responses, ingrained habits—these don't dissolve in the presence of love. They persist despite love.

Your partner might love you completely and still not be able to give you what you need. Your parent might care about you genuinely and still behave in ways that damage the relationship. Love is necessary for healthy relationships but not sufficient to fix everything broken.

"Love doesn't guarantee change" separates two things that feel like they should go together. It's possible to be loved and still not get the change you need. It's possible for someone to be unable to change even when they desperately want to.

This is painful because the alternative story is more comfortable. If they would change if they loved you, then the solution is clear: more love, or finding someone who loves you more. But if love and change are independent, then you have to grapple with a harder question: Can I accept this person as they are? Can I stay in this relationship without the change I want?

SHIFT 38

> **FROM:** *"I can fix them."*
>
> **TO:** *"I can only influence; they choose."*

This shift directly challenges the control fallacy.

"I can fix them" positions you as the agent of their change. Their problems are solvable—by you. If you say the right things, provide the right support, apply the right approach, they'll transform. The project is your responsibility.

But you can't fix people. You're not their therapist, their higher power, or their savior. You're a person in relationship with them, which gives you influence—not control.

"I can only influence" acknowledges what's actually available to you. You can express how their behavior affects you. You can request changes. You can set boundaries and consequences. You can model healthier patterns. You can make clear what you need. All of this is influence.

"They choose" locates agency where it belongs. After all your influence, the decision is theirs. They might change. They might not. That choice is made inside their head, in a place you cannot reach and cannot control.

This shift is liberating because it releases you from an impossible task. You're not failing to fix them—fixing them was never your job. Your job is to manage your own life, including deciding how to respond to someone who isn't changing despite your influence.

• • •

Putting It Together

When someone's behavior frustrates or hurts you, you have a sequence available.

Recognize the pattern. This isn't the first time. You've seen this behavior before. Name it: "They're doing that thing they do."

Apply reattribution. What explanations besides "they're trying to hurt me" might fit? What's going on in their world that might drive this behavior?

Accept their autonomy. You cannot make them change. You have never been able to. Their choices are theirs. Release the illusion of control.

Focus on your response. The one thing you control is how you respond. What boundaries serve you here? What action can you take? How will you handle this pattern going forward?

Decide what you're willing to accept. Accepting that you can't change them doesn't mean accepting any treatment. It means clearly seeing who they are and deciding whether you can live with that.

• • •

TAKE ACTION: YOUR NEXT STEPS

Reflection

Think of someone whose behavior consistently frustrates you—a family member, partner, friend, or colleague. Work through these questions:

1. What's the behavior pattern? Describe what they repeatedly do that bothers you. Be specific.

2. What have you tried? List the approaches you've used to change this behavior. Talking, boundaries, consequences, distance—what's been your strategy?

3. What explanations besides malice might apply? Practice reattribution. Why else might they behave this way? Fear? Insecurity? Habit? Limitation? Trauma?

4. What would accepting their pattern look like? If this behavior continues unchanged, what would your life look like? What adjustments would you need to make?

5. What do you actually control here? List the things you can actually do—responses, boundaries, decisions about the relationship. Focus only on what's in your domain.

Implementation

Practice the "that's what they do" phrase. The next time someone does that frustrating thing, say to yourself: "That's what they do." Not with resignation, but with clarity. Notice how naming the pattern changes your reaction.

Apply reattribution in real time. When someone's behavior triggers you this week, pause and generate three explanations besides "they're targeting me." Write them down. Notice if the emotional charge shifts.

Make one response change. Identify one aspect of your response to this person's behavior that you can change. Not their behavior—yours. Implement it and observe the results.

Troubleshooting

"But their behavior really is about me. They really are targeting me."

Maybe. Some people do target others deliberately. But even if the behavior is targeted, what changes? You still can't control their choices. The shifts still apply: focus on your response, release the illusion of control, decide what you're willing to accept. Whether the behavior is malicious or not, your options remain the same.

"This sounds like just accepting bad behavior. Where's the accountability?"

Accountability isn't control. You can hold someone accountable—communicate consequences, enforce boundaries, even end relationships—without controlling their choices. The shifts aren't about tolerating mistreatment. They're about locating your energy where it can actually work: in your own decisions and responses.

"What if the person is genuinely harmful? Shouldn't I try to protect others from them?"

If someone is dangerous, protect yourself and others as needed. These shifts aren't about enabling harm. They're about the psychological pattern of trying to change people through frustration and force of will. You can take protective action while accepting that you don't control their internal transformation.

• • •

Other people will do things you don't like. They'll repeat patterns that hurt you. They'll fail to change despite your best efforts to reach them. This is not a failure on your part—it's a feature of reality.

The shifts in this chapter offer freedom from the exhausting project of controlling others. You can want them to change. You can work to influence them. But the outcome is theirs to determine. And in releasing the demand that they be different, you reclaim energy for the one project you can actually complete: becoming who you want to be, regardless of what they do.

In the next chapter, we turn the lens inward. The patterns that frustrate you in others often echo patterns in yourself. The cognitive shifts for examining your own relational behavior—without self-attack or denial—offer something different: the possibility of actual change, because unlike with others, your choices genuinely are yours to make.

CHAPTER 11: Reframing Your Own Relational Patterns

Seeing your contribution clearly without self-attack

> *"When the light has been removed and my wife has fallen silent, aware of this habit that's now mine, I examine my entire day and go back over what I've done and said, hiding nothing from myself, passing nothing by. I have nothing to fear from my errors when I can say: 'See that you do not do this anymore. For the moment, I excuse you.'"*
>
> — *Seneca, On Anger*

You're lying awake at 2 AM, replaying the conversation from dinner.

You got defensive again. Your partner mentioned feeling neglected, and instead of listening, you launched into a list of everything you'd done for the family that week. You saw the hurt flash across their face. You heard your own voice getting sharper. And still you kept going.

Now the room is dark and quiet, and the verdict arrives: *I'm terrible at relationships. I always do this. I push people away. There's something wrong with me.*

The self-assessment feels honest. Even necessary. After all, the previous chapter was about releasing the delusion that you can control others. The natural next step is to look at yourself—the one person you actually can change. That's growth, right?

Not quite.

What you're doing isn't self-examination. It's self-attack. And the difference matters enormously—because one leads to change and the other perpetuates the very patterns you're trying to escape.

• • •

Labeling Versus Describing

Here's what happens when you examine your relational patterns without the right cognitive tools: you turn observations into identities.

"I got defensive in that conversation" becomes "I'm a defensive person."

"I wasn't fully present with my kids today" becomes "I'm a bad parent."

"I interrupted my friend three times" becomes "I'm selfish."

This is the cognitive distortion psychologists call *labeling*—taking a specific behavior in a specific context and inflating it into a global judgment about who you are. It feels like accountability. It's actually a trap.

When you label yourself, you create a fixed identity that feels permanent. The brain doesn't argue with identity statements. It simply files them as facts. And once "I'm bad at relationships" becomes a fact, change feels impossible. Why would you try to communicate differently when the problem is *who you are*?

Describing behavior works differently. "I got defensive when my partner raised that issue" is limited. It's one moment, one reaction, one context. It implies the possibility of a different reaction in a different moment. It leaves room for the question: *What could I do next time?*

The difference is subtle but everything depends on it:

Labeling: "I'm terrible at listening."

Describing: "I interrupted twice in that conversation."

Labeling: "I'm emotionally unavailable."

Describing: "I withdrew when she brought up her frustration."

Labeling: "I always make things worse."

Describing: "I escalated instead of de-escalating yesterday."

Labels close doors. Descriptions open them. If you want to actually change your relational patterns, you have to describe what happened without concluding what it means about your permanent character.

• • •

The Observer Behind the Pattern

Acceptance and Commitment Therapy introduces a concept that sounds abstract but proves revolutionary when you apply it: *self-as-context.*

The idea is this: you are not your thoughts, your feelings, or your patterns. You are the consciousness *observing* them.

Think of it as the difference between the weather and the sky. Your thoughts and behaviors are weather—sometimes sunny, sometimes stormy, constantly changing. But you are the sky—the unchanging space in which the weather appears.

When you say "I'm defensive," you're identifying with the weather. You've merged with a pattern and decided it defines you. But that identification isn't accurate. There's a part of you right now noticing your defensiveness, evaluating it, wondering how to change it. That observer isn't defensive. That observer is neutral—curious, even.

This isn't just philosophical wordplay. It has practical implications.

When you fuse with a pattern ("I AM defensive"), change feels like trying to become a different person. Impossible. When you observe a pattern ("I notice I have a defensive reaction"), change means adjusting behavior. Doable.

ACT calls the ability to step back and notice your own thoughts *defusion*. Instead of "I'm terrible at relationships," defusion sounds like: "I'm having the thought that I'm terrible at relationships." It's the same content, but now you're watching it rather than being consumed by it.

This distinction transforms self-examination. You can look honestly at your patterns—really see them, without minimizing—while simultaneously recognizing that you are not those patterns. You are the one who can choose to do something different.

• • •

The Stoic Evening Review

The ancient Stoics understood something modern psychology has confirmed: regular self-examination improves behavior, but only when done correctly.

Seneca, the Roman philosopher, had a nightly practice. After his wife fell asleep, he would mentally review his entire day—every conversation, every decision, every reaction. He describes this as "pleading my case at my own court."

But notice something crucial in his approach. After cataloging his errors, he says: "See that you do not do this anymore. *For the moment, I excuse you.*"

For the moment, I excuse you.

This isn't self-indulgence. This is strategy. Seneca understood that self-condemnation makes change less likely, not more. Shame doesn't motivate improvement—it triggers avoidance, denial, and repetition of the same patterns. You can't learn from something you're too busy hating yourself for.

The Stoic evening review has three elements worth adopting:

First, it's regular. You don't wait until a relationship crisis forces you to examine yourself. You do it nightly, systematically, when stakes are low and defensiveness is minimal.

Second, it's specific. You don't ruminate vaguely on your inadequacy. You ask concrete questions: What did I say? How did I react? What triggered me? What could I have done differently?

Third, it's compassionate. Not in a soft, excusing way—you hide nothing from yourself. But once you've noted the error and identified the correction, you release the judgment. "For the moment, I excuse you." Tomorrow is for doing better. Tonight is for understanding, not punishment.

The research supports this approach. Studies on self-compassion show that people who treat themselves kindly after failures are more likely to take responsibility and make changes than people who beat themselves up. Harsh self-criticism triggers defensive reactions—rationalization, blame-shifting, avoidance. Gentle self-examination allows you to face the truth because the truth is bearable.

• • •

SHIFT 39

FROM: *"I always do this"*
TO: *"I did this here; I can do differently next time"*

The phrase "I always" is a cognitive trap. It transforms a single instance into an unchangeable destiny.

"I always get defensive."

"I always shut down."

"I always push people away."

These statements feel true because pattern recognition is one of the brain's primary functions. You notice you did something problematic, scan your memory for similar instances, find them (because they exist—you're human), and conclude you have an unbreakable pattern.

But "I always" is almost never accurate. There were times you didn't get defensive. There were conversations where you stayed present. There were moments you let someone in. Those moments count, but the brain's negativity bias makes them harder to recall.

The reframe limits the claim to what's verifiable: "I did this here; I can do differently next time."

This isn't positive thinking or denial. You're not pretending the pattern doesn't exist. You're accurately locating it in time and space—this conversation, this moment, this context. And you're adding something the "always" statement leaves out: your capacity to choose differently.

The shift acknowledges both realities: yes, you have a pattern. No, you are not enslaved to it. Every interaction is a new opportunity. The past informs but doesn't determine the future.

• • •

SHIFT 40

FROM: *"I'm the problem"*

TO: *"I contributed; so did the situation and other people"*

When relationships struggle, there's a temptation to assign blame in clean, complete packages. Either the other person is entirely at fault, or you are.

Neither is accurate. Relational dynamics are always co-created.

Consider the conversation you replay at night. Yes, you got defensive. But what was the context? Was the topic sensitive? Was the timing terrible? Had previous conversations on this subject gone badly? Was there something in the other person's tone or framing that activated your defense mechanisms?

None of this excuses your reaction. But it contextualizes it. And context matters because it reveals intervention points you might otherwise miss.

If you conclude "I'm the problem," the only solution is to somehow become a different person. Overwhelming and vague. If you conclude "I contributed, alongside other factors," you can identify what specifically to change. Maybe it's the reaction itself. Maybe it's having the conversation at a different time. Maybe it's asking for a different approach from the other person. Multiple levers, multiple possibilities.

This shift also protects against the flip side: concluding the other person is entirely at fault. Both extremes are seductive because they're simple. Complexity is harder to hold but closer to truth.

You contributed. So did they. So did the circumstances. Now—which part of your contribution do you want to work on?

• • •

SHIFT 41

> **FROM:** *"I'm bad at relationships"*
>
> **TO:** *"I have patterns that aren't serving me"*

This shift is the labeling-versus-describing distinction applied to your relational identity.

"I'm bad at relationships" is a death sentence. It implies a fixed incapacity—something broken at your core that can't be repaired. It explains every past failure and predicts every future one. And because it feels so final, it actually discourages effort. Why try when you're constitutionally unable?

"I have patterns that aren't serving me" describes the same reality but changes the relationship to it.

Notice what the reframe does:

It's specific. Not a global judgment about your relational capacity but an observation about particular behaviors.

It's temporary. Patterns can change. They're learned responses, not genetic destiny.

It's functional. "Not serving me" implies a standard—these behaviors aren't moving you toward what you want. That invites the question: what patterns would serve you?

The man who says "I'm bad at relationships" stops examining. He knows the diagnosis. The man who says "I have patterns that aren't serving me" starts analyzing. Which patterns? Where did they come from? What triggers them? What would different patterns look like?

Same information. Different cognitive frame. Different outcomes.

• • •

SHIFT 42

> **FROM:** *"It's too late to change"*
>
> **TO:** *"It's only too late if I stop here"*

This is the shift against despair.

After years of the same patterns—the same defensive reactions, the same withdrawal, the same communication failures—it starts to feel calcified. You've been this way so long. You've hurt people with these patterns for decades. What's the point of changing now?

The sunk cost fallacy in reverse: because so much damage has already been done, further effort feels useless.

But "too late" is almost never true in relational terms. Relationships are live. People you've hurt are often still in your life, waiting to see if you'll change. Conversations you handled badly can be revisited, repaired, approached differently next time.

More importantly, the patterns themselves don't have an expiration date for modification. Research on neuroplasticity shows that the brain remains changeable across the lifespan. New responses can be learned at forty, at sixty, at eighty. The idea that personality is set in stone by early adulthood has been thoroughly disproven.

What is true: change gets harder with time because patterns deepen. Each repetition strengthens the neural pathway. The defensive reaction that was a choice at twenty feels automatic at fifty.

But "harder" isn't "impossible." And the alternative—giving up—guarantees the very outcome you fear. If you believe it's too late and stop trying, then yes, nothing will change. Self-fulfilling prophecy.

The reframe is simpler and more honest: "It's only too late if I stop here." Change is possible in this moment, regardless of how many moments came before.

• • •

Compassionate Self-Examination in Practice

These four shifts share a common architecture: they allow you to see your contribution clearly without collapsing into self-attack.

This is harder than it sounds. Most people default to one extreme or the other. They either minimize their role (it's not my fault, they made me react that way) or they catastrophize it (I'm the problem, I always ruin everything, it's too late to change).

The middle path requires holding two things simultaneously:

Yes, I did something that didn't work.

No, this doesn't mean I'm fundamentally broken.

Compassionate self-examination isn't soft. You look directly at what happened. You name your contribution without rationalization. You trace the pattern to its roots. None of this is comfortable.

But the judgment you apply is different. Instead of "I'm terrible," you ask "What can I learn?" Instead of "I always do this," you ask "What triggered this instance?" Instead of "It's too late," you ask "What would I do differently tomorrow?"

This is the stance Seneca was pointing toward. Hide nothing. Pass nothing by. And then: "See that you do not do this anymore. For the moment, I excuse you."

• • •

Putting This Into Practice

TAKE ACTION: YOUR NEXT STEPS

Reflection Questions

1. What is a relationship pattern you've labeled as a permanent character trait? How would you describe the same pattern as a specific, changeable behavior?

2. Think of a recent relational conflict. What was your contribution? What else contributed (other people, timing, context, history)?

3. When you examine your own behavior, do you tend toward minimizing or catastrophizing? What does your inner critic typically say?

Action Steps

1. Implement a Stoic evening review. Before sleep, mentally review one conversation from the day. What did you say? How did you react? What worked? What didn't? End with: "Tomorrow I will do better. For now, I excuse myself."

2. Catch labeling in action. When you notice a self-judgment that sounds like identity ("I'm selfish," "I'm a terrible partner"), immediately translate it to behavior ("I made a selfish choice in that moment," "I handled that badly").

3. Practice defusion. When harsh self-criticism appears, precede it with "I'm having the thought that..." Notice how this creates distance between you and the thought.

4. Replace "always" with "here." Monitor your language. When you catch yourself saying "I always do this," correct to "I did this here." Feel the difference in what becomes possible.

5. Complete a contribution analysis. Take a recent conflict and divide a piece of paper into three columns: "My Contribution," "Their Contribution," "Situational Factors." Fill in all three honestly. Notice that the first column is the only one you can change.

Troubleshooting

"But what if I really am the problem? What if it's always my fault?"

It's not. Relational dynamics are always co-created. But let's assume for a moment you contribute more than others to conflicts in your life. The shifts still apply. You can acknowledge your significant contribution without defining yourself by it. You can have patterns that cause repeated problems without concluding you're fundamentally broken. The description-versus-label distinction matters most precisely when there's genuine accountability to be taken.

"This sounds like letting myself off the hook."

The opposite. When you attack yourself, you trigger defensiveness—even against yourself. Harsh judgment makes you want to avoid the topic, rationalize, or give up. Compassionate examination allows you to stay with the discomfort long enough to actually learn from it. The goal isn't to feel better about your mistakes. The goal is to change your behavior. Self-compassion serves that goal better than self-attack.

"What if I've already tried to change and failed?"

Then you tried an approach that didn't work. That's data, not destiny. Most lasting change requires multiple attempts with different strategies. The fact that previous efforts failed doesn't mean future efforts will. It means you need a different method—perhaps the one outlined here, perhaps something else. "It's only too late if I stop here."

• • •

The patterns that frustrate you in yourself can be changed. Not by hating them, not by denying them, but by seeing them clearly and choosing to do something different.

That's the opportunity you have that you don't have with others. Their patterns are theirs to change or not. Yours are yours—not your permanent identity, but your current behavioral defaults. Defaults can be overwritten.

The next chapter turns to the deepest and most vulnerable layer of relational life: connection and intimacy themselves. The cognitive shifts that govern how close you're willing to let people get, and what you believe you deserve from love.

CHAPTER 12: Reframing Connection and Intimacy

The shifts that make closeness possible

"Vulnerability is not weakness. And that myth is profoundly dangerous."

— Brené Brown

You want to tell your partner something important. Not logistical—emotionally important. Something you've been carrying alone. Maybe it's a fear about the future. Maybe it's an old wound that's affecting how you show up in the relationship. Maybe it's just that you need more from them than you've been asking for.

The words form in your mind. You take a breath. And then—nothing. The moment passes. You say something else instead. Something safer. The real thing stays locked inside, where it's been for months. Maybe years.

Later, alone, you tell yourself a story about why: *They wouldn't understand anyway. I shouldn't need to explain this. If they really knew me, they'd already get it. Besides, it's not that big a deal.*

None of that is true. But it feels true, and that's the problem.

• • •

Emotional Reasoning in Relationships

Aaron Beck, the founder of Cognitive Behavioral Therapy, identified a cognitive distortion he called *emotional reasoning*: mistaking feelings for facts. When we feel anxious, we assume something dangerous is happening. When we feel sad, we assume our situation is hopeless. The emotion becomes evidence.

In relationships, emotional reasoning is particularly destructive because it governs whether we move toward connection or away from it.

Consider vulnerability. When you contemplate sharing something real—a fear, a need, an imperfection—you feel exposed. Maybe scared. The emotional reasoning kicks in: *This feeling of danger means sharing is actually dangerous. This discomfort is a warning sign. I should protect myself.*

But the feeling isn't evidence. The feeling is just a feeling. It might be telling you something important, or it might be a conditioned response from times when vulnerability actually was dangerous. Your nervous system doesn't distinguish between genuine threats and echoes of old wounds.

The result: you treat a feeling like a fact and act accordingly. You stay quiet when speaking would serve you. You keep distance when closeness is what you want. You protect yourself from something that isn't actually threatening—and in doing so, you lose what you most deeply need.

• • •

The Avoidance Trap

Here's how the cycle works:

You value connection. You want intimacy. But intimacy requires vulnerability, and vulnerability triggers discomfort. So you avoid the vulnerability. In the short term, this works—the discomfort goes away. But in the long term, you've moved further from what you value, not closer.

Acceptance and Commitment Therapy calls this *experiential avoidance*: the strategy of organizing your behavior around avoiding uncomfortable internal experiences. It makes sense as a short-term tactic. It's catastrophic as a life strategy.

The problem is that discomfort is the price of admission for nearly everything meaningful. Connection requires vulnerability, which feels exposed. Growth requires risk, which feels scary. Love requires trust, which feels uncertain. If you only do things that feel comfortable, you're confined to a shrinking territory that excludes most of what makes life worth living.

ACT offers a radical alternative: *willingness*. Not the absence of discomfort, but the choice to experience discomfort in service of something you value. You don't have to like the feeling. You don't have to believe it will work. You just have to be willing to have the feeling while doing what matters anyway.

This is what makes the cognitive shifts in this chapter different from mere positive thinking. You're not convincing yourself that vulnerability is easy or that discomfort doesn't exist. You're recognizing that the discomfort is real and choosing to act in spite of it—because the alternative is a life organized around avoidance, which is no life at all.

• • •

SHIFT 43

FROM: *"Vulnerability is weakness"*

TO: *"Vulnerability is the price of connection"*

The story you've absorbed about vulnerability is probably wrong.

Most cultures frame vulnerability as something to overcome—a deficiency to be managed, a gap in your armor that enemies could exploit. The messaging is consistent: strong people don't need to expose themselves. Revealing your inner world is risky and weak.

This framing misses something fundamental: vulnerability isn't the opposite of strength. It's the prerequisite for connection.

Brené Brown's research on this is unambiguous. After twelve years studying human connection, she found that people who experienced deep belonging shared a common trait: they were willing to be seen. Not the curated version. Not the highlight reel. The actual person, including the parts that felt risky to reveal.

Her definition of vulnerability is precise: "uncertainty, risk, and emotional exposure." Notice there's nothing passive or weak about this. Vulnerability is the act of showing up when you can't control the outcome. That requires more courage, not less.

The reframe acknowledges reality: vulnerability isn't inherently good or bad. It's a *cost*. The question is whether the cost is worth paying. And for connection—real connection, not its performance—vulnerability isn't optional. It's the admission fee.

You can't be truly known while remaining hidden. You can't experience intimacy while maintaining armor. The exposure is the point.

This doesn't mean indiscriminate disclosure. Vulnerability should be earned—shared with people who have demonstrated they can hold it. But within relationships where connection is the goal, the unwillingness to be vulnerable is the unwillingness to connect. The two are inseparable.

• • •

SHIFT 44

FROM: *"I shouldn't need anyone"*

TO: *"Interdependence is strength, not deficiency"*

The hyper-individualist narrative runs deep: needing others is a failure of self-sufficiency. Real strength means handling everything yourself. Asking for help reveals weakness.

This narrative is both culturally dominant and psychologically backward.

Humans are biologically wired for connection. We evolved as tribal animals because cooperation dramatically increased survival. Our nervous systems are designed to co-regulate with others—to be calmed by presence, stressed by isolation. The research on loneliness shows it's as damaging to health as smoking.

The idea that needing connection is a deficiency ignores what we are.

Dependence—unhealthy reliance that diminishes your functioning—is different from interdependence. Dependence says: I can't function without you. Interdependence says: I function better with you, and you function better with me.

Healthy interdependence includes the ability to ask for what you need. To lean on others during hard times. To accept support without shame. These aren't admissions of inadequacy. They're recognitions of how humans work.

The "I shouldn't need anyone" frame often masks fear. If you don't need anyone, you can't be hurt when they disappoint you. If you handle everything alone, you never have to experience the vulnerability of asking. The independence becomes armor—effective at blocking pain but equally effective at blocking connection.

The shift recognizes something important: strength isn't the absence of need. It's the capacity to meet needs skillfully—including the capacity to receive from others. Relationships require both giving and receiving. Someone who only gives isn't being generous; they're maintaining control while avoiding the vulnerability of need.

Interdependence is how humans thrive. Not weakness. Design.

• • •

SHIFT 45

FROM: *"They should just understand me"*

TO: *"I am responsible for expressing myself clearly"*

This shift addresses a belief that sabotages countless relationships: the idea that people who truly love us should intuit our inner world without being told.

If they really cared, they'd know. I shouldn't have to explain—they should just get it. The fact that I have to ask proves they don't understand me.

These thoughts feel true because they're emotionally seductive. There's something romantic about being understood without words. And there's something vulnerable about having to articulate what you need—it admits that you have needs, that you're not self-sufficient, that you require something from another person.

But the belief is a trap. It sets an impossible standard. Then it punishes your partner for failing to meet it.

No one can read minds. Even in the closest relationships, even between people who know each other deeply, assumptions about internal states are often wrong. You experience your thoughts, feelings, and needs in high definition. Others experience only what you communicate—and often less than you think.

The shift places responsibility where it belongs: on you, for expressing yourself clearly. Not because your partner doesn't care. Not because the relationship is flawed. But because communication is how two separate consciousnesses actually connect. There's no other way.

This reframe has a liberating implication. When you accept responsibility for expressing yourself, you gain power over the outcome. You're no longer waiting passively to be understood. You're actively creating understanding by saying what you mean, asking for what you need, and clarifying when you're misinterpreted.

The alternative—waiting to be understood without speaking—is a recipe for resentment. You suffer in silence while your partner remains unaware. Then you resent them for not knowing what they couldn't possibly know. The cycle repeats until connection erodes entirely.

Being responsible for clear expression isn't a burden. It's the only path to being known.

• • •

Values-Consistent Action When Uncomfortable

The three shifts in this chapter share a common thread: they all require action that feels uncomfortable.

Being vulnerable feels exposed. Accepting interdependence feels like admitting inadequacy. Expressing yourself clearly feels risky. If you wait until these things feel comfortable, you'll wait forever.

ACT's answer is *committed action*: behavior that aligns with your values even when—especially when—difficult thoughts and feelings show up. Values provide the reason to move forward. Willingness provides the capacity.

Here's what this looks like in practice:

You value intimacy with your partner. To have intimacy, you need to share something real about yourself. Sharing triggers fear and the thought "what if they reject me?" The fear is present. The thought is present. And you share anyway—not because you've conquered the fear, but because you're willing to experience it in service of what you value.

This is fundamentally different from waiting to feel ready. Readiness rarely arrives on its own. It emerges from action. You become someone who can be vulnerable by practicing vulnerability. You become someone who can ask for what you need by asking. The skill develops through repetition, not preparation.

The emotions may not change. After years of practice, vulnerability might still feel scary. But your relationship to the fear changes. It becomes something you carry with you rather than something that stops you. Background noise rather than a veto.

• • •

What Intimacy Actually Requires

Let's be concrete about what these shifts demand:

Intimacy requires being seen in ways that feel uncomfortable. Not just the flattering parts—the parts you'd rather hide. Your fears. Your failures. Your needs. The gap between who you pretend to be and who you actually are.

Intimacy requires receiving as well as giving. Allowing others to support you. Accepting care without immediately returning it. Sitting with the vulnerability of being held.

Intimacy requires articulating your inner world clearly. Not dropping hints. Not testing whether they'll figure it out. Actually saying what you mean, what you feel, what you need—even when you'd prefer they already knew.

None of this is easy. But consider the alternative: relationships where you're never truly seen, where giving is a way to avoid receiving, where you resent people for not knowing things you never told them. This is protection that prevents exactly what it claims to want.

The shifts in this chapter open a door. Walking through it requires willingness to be uncomfortable. But on the other side is something that avoidance can never provide: the experience of being truly known and loved, not despite your real self, but because of it.

• • •

Putting This Into Practice

TAKE ACTION: YOUR NEXT STEPS

Reflection Questions

1. What do you avoid expressing in your closest relationships? What would become possible if you said it?

2. Where in your life have you mistaken independence for strength, avoiding the vulnerability of asking for help?

3. Think of a recent moment when you resented someone for "not understanding." Did you actually tell them what you needed? If not, what stopped you?

Action Steps

1. **Identify one thing you've been withholding.** Something real—a need, a fear, a feeling you've kept private from someone close to you. Commit to sharing it within the next week. Notice the discomfort beforehand, share anyway, and observe what actually happens versus what you feared.

2. **Practice receiving.** The next time someone offers help, accept it without deflecting or immediately reciprocating. Stay with the discomfort of being supported. Notice the urge to re-establish self-sufficiency, and resist it.

3. **Replace hints with direct requests.** When you need something from someone, say it explicitly. "I need..." or "Would you..." rather than hoping they'll infer it. Track how much more effective direct communication is compared to waiting to be understood.

4. **Name the emotional reasoning.** When you feel scared about being vulnerable, notice the thought: "This feeling means it's dangerous." Then ask: Is the feeling evidence, or is the feeling just a feeling? What would you do if you

acted on your values rather than your fear?

5. **Set a values-based goal.** Choose one relationship where you want more connection. Identify one values-consistent action—something that requires vulnerability—and commit to it. Write down the uncomfortable feelings you're willing to experience in order to take this action.

Troubleshooting

"What if I'm vulnerable and they actually do reject me?"

Then you'll have important information. Not everyone can hold what you share. Not every relationship can bear the weight of intimacy. Finding out who can and can't is valuable, even when painful. The alternative—never knowing because you never risked—keeps you in relationships where you're never truly seen. That's a slower kind of rejection.

"I've been hurt before when I was vulnerable. Isn't self-protection reasonable?"

Self-protection is reasonable when the threat is real. The question is whether you're protecting yourself from actual danger or from the memory of past danger. If every relationship gets punished for wounds inflicted by previous ones, no relationship can succeed. Past hurt explains caution. It doesn't justify permanent withdrawal.

"What if I express myself clearly and they still don't give me what I need?"

Then you have a different problem than the one addressed by this shift. If you've expressed yourself clearly and they consistently can't or won't meet your needs, that's information about compatibility or willingness—not a communication failure. But most people never reach this clarity because they never express themselves directly enough to find out.

• • •

The deepest relationships require the greatest risk. Connection at its core is the experience of being seen and accepted, which requires first being seen—and that requires showing yourself.

The shifts in this chapter won't eliminate the discomfort of vulnerability. They won't make independence unnecessary or communication effortless. What they offer is a clearer view of the trade-off: you can have safety or you can have intimacy, but you can't have both without cost.

Choose your values. Accept the cost. And do what matters anyway.

This completes Part Three. The relationship shifts—from conflict to other people's behavior to your own patterns to connection itself—cover the territory where most suffering and most flourishing occur. What remains is resilience: how you frame setbacks, uncertainty, and the inevitable difficulties of life.

PART FOUR

RESILIENCE SHIFTS

CHAPTER 13: Reframing Adversity and Hardship

The cognitive shifts that transform obstacles into material

> *"The impediment to action advances action. What stands in the way becomes the way."*
>
> — Marcus Aurelius, *Meditations*

The diagnosis arrives. The layoff comes. The relationship ends. The market crashes. The thing you built fails. The person you trusted betrays you.

In the moment, the mind does what minds do: it recoils. It protests. It generates a cascade of thoughts designed to reject what has happened.

This can't be happening. This shouldn't be happening. This is the worst thing that could happen.

These thoughts feel appropriate—even righteous. Reality has violated your expectations, and your mind is filing a formal complaint. But here's the problem: the complaint changes nothing. The thing has already happened. And while you're busy arguing with reality, reality is moving forward without you.

The question isn't whether adversity is fair. The question is what you do now that it's here.

• • •

The Stoic Discipline of Assent

Nearly two thousand years ago, the Stoic philosopher Epictetus taught his students a practice he called the *discipline of assent*. The idea is deceptively simple: between an event and your reaction to it, there is a gap. In that gap, you have a choice about what meaning to assign.

Events generate *impressions*—raw data about what has happened. But impressions are not judgments. The layoff is an impression. "This is catastrophic" is a judgment you add afterward.

Epictetus put it directly: "Be not swept off your feet by the vividness of the impression, but say, 'Impression, wait for me a little. Let me see what you are and what you represent. Let me test you.'"

This pause—this testing—is where resilience lives. The Stoics understood that we don't control what happens to us, but we control what we assent to. We control which interpretations we accept and which we reject.

The discipline of assent doesn't ask you to pretend bad things aren't bad. It asks you to separate the fact from the story. The job loss is a fact. "My career is over" is a story. The diagnosis is a fact. "This is the end of everything good" is a story.

Modern cognitive behavioral therapy rediscovered this insight two millennia later. Aaron Beck identified *catastrophizing* as a cognitive distortion—the tendency to assume the worst possible outcome and treat it as certain. The antidote, which CBT calls *decatastrophizing*, involves examining whether your catastrophic interpretation is actually warranted by the evidence.

The Stoics and the cognitive therapists arrived at the same conclusion by different routes: what devastates you isn't usually the event itself. It's the interpretation you reflexively attach to it.

• • •

SHIFT 46

FROM: *"This shouldn't be happening to me"*

TO: *"This is happening; what's possible now?"*

The phrase "this shouldn't be happening" contains a hidden assumption: that reality owes you something different than what it's delivering. That you have a contract with the universe, and the universe is in breach.

This assumption is understandable. When you've worked hard, made good decisions, and treated people well, adversity feels like a violation of natural law. You followed the rules. Why isn't the universe following them back?

The problem is that the universe has no rules in this sense. It has patterns, tendencies, probabilities—but no guarantees. Bad things happen to careful people. Good decisions sometimes lead to bad outcomes. The world is not a vending machine where virtue goes in and reward comes out.

"This shouldn't be happening" is not merely inaccurate. It's also expensive. While you're consumed with the injustice of your situation, you're not responding to it. You're litigating with reality instead of adapting to it.

Marcus Aurelius, the Roman emperor and Stoic philosopher, faced this choice constantly. War, plague, betrayal, the death of children—his journals are filled with reminders to himself about how to think when reality delivers something unwanted. His most famous formulation: "The mind adapts and converts to its own purposes the obstacle to our acting. The impediment to action advances action. What stands in the way becomes the way."

This isn't optimism. It's strategy. Marcus isn't claiming that obstacles are secretly good or that adversity is a blessing in disguise. He's making a practical observation: since you cannot remove the obstacle, your only option is to work with it. And working with it often reveals possibilities that weren't visible before.

The shift from "shouldn't be happening" to "is happening; what's possible now?" doesn't require you to like what's happening. It requires you to accept that it's happening—and then to redirect your energy from protest to response.

Acceptance, in this sense, is not approval. It's acknowledgment. You can hate your situation and still accept that it's your situation. That acceptance is what frees your attention for the question that actually matters: given this reality, what can you do?

• • •

The Practice of Decatastrophizing

When adversity strikes, the mind races to the worst-case scenario. This is natural—your brain is trying to prepare you for danger. But the preparation often overshoots. What feels like clear-eyed realism is usually cognitive distortion.

Decatastrophizing is the practice of examining your catastrophic interpretations with the same rigor you'd apply to any other claim. It asks three questions:

First: *What is the actual evidence for my worst-case interpretation?* Not the feeling that it's true—the evidence. If you've been laid off and your mind says "I'll never work again," what actual evidence supports that? Have you applied to jobs and been rejected by every employer in your field? Or are you predicting a future that hasn't happened yet?

Second: *What's the realistic range of outcomes?* Catastrophizing collapses all possibilities into the single worst one. But outcomes distribute across a range. What's the best case? The most likely case? The worst case? Usually, the worst case is possible but not probable, and even the most likely case is better than what your catastrophizing mind assumes.

Third: *If the worst case did happen, could you cope with it?* Catastrophizing assumes not only that the worst will happen but that you'll be destroyed by it. But you've survived difficulty before. You've adapted to circumstances you never imagined you could handle. Your coping capacity is almost certainly greater than your catastrophizing mind gives you credit for.

This isn't about forcing positivity. It's about accuracy. Catastrophizing distorts reality in the direction of despair. Decatastrophizing corrects the distortion and shows you reality as it actually is—which is usually harder than you'd like but more workable than you feared.

• • •

SHIFT 47

FROM: *"This is the worst thing"*

TO: *"This is hard; I don't know yet what it will become"*

In the immediate aftermath of adversity, everything looks like disaster. This is partly because you're experiencing the acute phase of loss, and partly because you can't yet see how the story unfolds.

The frame "this is the worst thing" treats the present moment as the final verdict. It assumes you know exactly what this event means and that its meaning is entirely negative. But you don't know. You're too close. You're standing in the middle of a story whose ending hasn't been written.

Research on post-traumatic growth, pioneered by psychologists Richard Tedeschi and Lawrence Calhoun, reveals something that seems paradoxical: many people who experience significant adversity eventually report positive changes that they attribute to the struggle. Not despite the difficulty—because of it.

The five domains of post-traumatic growth tell the story: greater appreciation of life (priorities clarify, small pleasures matter more), improved relationships (connections deepen, compassion expands), increased sense of personal strength (you discover capacities you didn't know you had), recognition of new possibilities (doors open that were invisible before), and spiritual or existential development (meaning deepens, perspective widens).

This doesn't mean adversity is good or that you should seek it. It means that the relationship between hardship and outcome is more complex than the catastrophizing mind assumes. What looks like pure loss in the moment may—over months or years—become something more nuanced.

The shift from "this is the worst thing" to "this is hard; I don't know yet what it will become" preserves accuracy without abandoning hope. You acknowledge the genuine difficulty of your situation. You also acknowledge that you're not yet in a position to know its full meaning.

This is not denial. It's epistemic humility. You're admitting that your current vantage point is limited, that time may reveal dimensions you can't see yet, and that premature certainty about outcomes is a form of self-deception.

• • •

The Obstacle as Material

The Stoic practice of turning obstacles upside down isn't about pretending difficulties are opportunities. It's about recognizing that difficulties are always, at minimum, opportunities to practice virtue.

Someone blocks your path? An opportunity to practice patience. Plans fall apart? An opportunity to practice adaptability. You lose something you valued? An opportunity to practice detachment.

This may sound like cold comfort when you're in acute pain. And it is—cold comfort is still comfort. The point isn't to feel good about adversity. It's to prevent adversity from being pure loss. Even in the worst circumstances, you can still choose how you respond. That choice is a form of agency, and exercising it is a form of growth.

Ryan Holiday, drawing on Marcus Aurelius, describes this as the discipline of seeing obstacles not as roadblocks but as raw material. The sculptor doesn't resent the marble for being hard. The hardness is what makes sculpture possible. Similarly, the difficulty in your life isn't merely an obstruction to your goals—it's the material from which character is formed.

This reframe doesn't make hardship pleasant. What it does is make hardship meaningful. And meaning, research consistently shows, is one of the most powerful predictors of resilience. People who can construct meaning from difficulty recover better, adapt faster, and emerge more capable than those who experience difficulty as purely random and pointless.

• • •

What the Shifts Demand

The two shifts in this chapter—from "shouldn't be happening" to "what's possible now," and from "this is the worst thing" to "I don't know yet what it will become"—require something of you.

They require you to give up the satisfaction of outrage. There's a certain grim pleasure in being wronged, in having evidence that the universe is unfair. The shifts ask you to trade that pleasure for effectiveness.

They require you to tolerate uncertainty. The frame "I don't know yet" is uncomfortable precisely because you want to know. You want closure. You want to understand what this means so you can stop thinking about it. But premature closure is a lie you tell yourself. The shifts ask you to sit with not-knowing while you gather more information.

They require you to act before you feel ready. Acceptance doesn't arrive as a feeling you wait for. It arrives as a choice you make—often before you feel accepting. You decide to treat reality as the starting point, and the feeling catches up later.

None of this is easy. But consider the alternative: you stay stuck in protest, catastrophizing about outcomes you can't predict, paralyzed by interpretations that may not even be accurate. The difficulty of the shifts is real. The cost of avoiding them is higher.

• • •

Putting This Into Practice

Reflection Questions

1. Think of a current difficulty you're facing. What interpretation have you attached to it, and how much of that interpretation is inference versus established fact?

2. Recall a past hardship that seemed catastrophic at the time. How did its meaning change as time passed? What became possible that wasn't visible in the moment?

3. When you encounter obstacles, do you tend toward "this shouldn't be happening" or toward "what can I do with this"? What would shift if you changed your default?

Action Steps

1. **Practice the Stoic pause.** When something unwanted happens, before reacting, say to yourself: "Impression, wait. Let me see what you really are." Give

yourself ten seconds to separate the fact from the story you're adding to it.

2. **Decatastrophize in writing.** Take a current worry and write down: (a) the worst-case scenario you're imagining, (b) the actual evidence for that scenario, (c) the realistic range of possible outcomes, and (d) your capacity to cope even if the worst happened. Notice how the exercise changes your relationship to the fear.

3. **Adopt the "not yet" frame.** For the next week, whenever you catch yourself labeling something as "the worst" or "a disaster," add the phrase "I don't know yet what this will become." Track how this small addition affects your emotional state and your ability to respond.

4. **Identify the obstacle as material.** When facing a difficulty, ask: "What virtue can I practice here? What skill can I develop? What character trait is this situation demanding of me?" Treat the obstacle as a workout for qualities you want to strengthen.

5. **Review past obstacles.** Make a list of three significant difficulties from your past and note what, if anything, became possible because of them—relationships that formed, strengths you developed, directions you took. Use this as evidence that the story of your current difficulty isn't finished yet.

Troubleshooting

"This feels like toxic positivity—pretending bad things aren't bad."

These shifts don't ask you to pretend. They ask you to be accurate. Bad things are bad. Loss is painful. The question is whether your interpretation of the bad thing is accurate—whether you're seeing it clearly or adding distortions that make it seem worse and more permanent than it is. Decatastrophizing isn't optimism. It's precision.

"Some things really are catastrophic. Not everything has a silver lining."

Correct. Some losses are permanent. Some damage is irreversible. The shifts don't promise that everything will be fine. They promise that (a) you probably don't know yet whether this thing is permanently catastrophic, and (b) even if it is, your response to it remains within your control. Post-traumatic growth is not universal. But the capacity to respond with agency to even the worst circumstances is.

"I can't just accept something terrible. Acceptance feels like giving up."

Acceptance isn't giving up—it's starting from reality. You can accept that something has happened and still work to change its consequences. The person who accepts a diagnosis doesn't stop seeking treatment. The person who accepts a layoff doesn't stop looking for work. Acceptance frees energy that was being wasted on denial so it can be redirected toward action.

• • •

The next chapter extends the resilience theme to something equally difficult: uncertainty. If adversity is the challenge of dealing with what has happened, uncertainty is the challenge of dealing with what might happen. The same principles apply—but the application looks different when the threat is still hypothetical.

CHAPTER 14: Reframing Emotion and Reaction

The shifts that create space between trigger and response

> *"Between stimulus and response there is a space. In that space is our power to choose our response. In our response lies our growth and our freedom."*
>
> — Often attributed to Viktor Frankl

Your phone buzzes. You glance at the screen and see who's calling. Before you consciously think anything, something happens in your body. Your stomach tightens. Your jaw clenches. Your heart rate increases.

You haven't answered yet. You haven't heard what they want. But the emotion is already there—dread, irritation, anxiety—triggered by a name on a screen.

And here's the thing: the emotion feels like information. It feels like it's telling you something true. *This call will be bad. This person always causes problems. I can't deal with this right now.*

But the emotion isn't information in the sense you think. It's a signal, not a report. It's telling you something about your nervous system's prediction, not about what's actually happening or what you should do.

The question isn't whether you can stop the emotion from arising. You can't. The question is what happens next.

• • •

The Emotional Reasoning Distortion

Aaron Beck, the founder of Cognitive Behavioral Therapy, identified a pattern he called *emotional reasoning*: using feelings as evidence for conclusions. If I feel anxious, something must be dangerous. If I feel hopeless, things must be hopeless. The emotion becomes proof that the corresponding belief is true.

This logic seems intuitive because emotions do often track reality. You feel fear when there's danger. You feel sadness when there's loss. But the relationship isn't reliable enough to use emotions as your primary source of truth. Emotions are influenced by sleep, blood sugar, hormones, past trauma, conditioned associations, and dozens of other factors that have nothing to do with the present moment.

The person calling might have good news. The conversation might go better than expected. The thing you're dreading might not happen at all. But emotional reasoning doesn't check these possibilities. It takes the feeling at face value and treats it as evidence.

In relationships, emotional reasoning is particularly destructive. You feel hurt, so you conclude your partner intended to hurt you. You feel anxious about your friend's silence, so you conclude they're angry at you. You feel jealous, so you conclude something suspicious is happening. The emotion dictates the interpretation, regardless of what's actually occurring.

The problem isn't that emotions are wrong—they often aren't. The problem is that emotions are *unreliable* as sole evidence. They're one input, not the whole picture. Treating them as the whole picture leads you to conclusions you wouldn't endorse if you examined them more carefully.

• • •

SHIFT 48

FROM: *"I can't help how I feel"*

TO: *"I can't choose the feeling; I can choose the action"*

This shift acknowledges something important: you really can't choose your emotions in the moment they arise. The anxiety hits before you decide to be anxious. The anger flares before you decide to be angry. Emotions are involuntary responses—products of conditioning, chemistry, and context that precede conscious thought.

But there's a difference between the feeling and what you do with it.

You feel anxious, and you can avoid the conversation or have it anyway. You feel angry, and you can lash out or pause. You feel hurt, and you can withdraw or express what you need. The feeling doesn't dictate the action. The feeling creates an urge, an impulse, a pull in a particular direction—but between the feeling and the behavior, you retain a choice.

This distinction matters because "I can't help how I feel" often slides into "I can't help what I do." The feeling becomes the excuse for the action. *I'm sorry I yelled—I was angry. I couldn't help withdrawing—I felt hurt. I had to check their phone—I was jealous.*

The shift reclaims agency without denying reality. Yes, the feeling arose without your consent. No, that doesn't mean your response is automatic. You can feel afraid and still move forward. You can feel angry and still speak calmly. You can feel jealous and still choose trust. The feeling is real; the action is yours.

This is uncomfortable because it removes a convenient defense. You can no longer point to your emotions as though they operated you like a puppet. But it's also liberating because it means you're not at the mercy of whatever you happen to feel. You can have emotions without being had by them.

• • •

The Passengers on the Bus

Acceptance and Commitment Therapy offers a metaphor that captures this relationship between emotions and actions: passengers on a bus.

Imagine your life is a bus, and you're the driver. The direction of the bus represents your values—where you actually want to go. But you're not alone on the bus. Various passengers ride with you: thoughts, feelings, memories, urges. Some passengers are pleasant. Others are loud, critical, threatening. They shout directions at you, insist you're going the wrong way, demand that you let them drive.

When you try to turn toward something that matters to you—a new direction, a meaningful goal, a difficult conversation—the unpleasant passengers get louder. They stand up in the aisle. They yell that you can't handle this, that you should turn back, that something terrible will happen if you continue.

Here's where people typically make one of two mistakes.

The first mistake is fighting the passengers. You stop the bus and try to throw them off. You argue with them, wrestle with them, try to convince them they're wrong. The problem: while you're fighting with the passengers, the bus isn't moving. You're not going anywhere because all your energy is consumed by the internal battle.

The second mistake is obeying the passengers. You let them dictate your route. If they say turn around, you turn around. If they say avoid that road, you avoid it. The problem: now the passengers are driving, not you. The direction of your life is determined by whatever emotional noise happens to be loudest.

The alternative is to keep driving. The passengers can stay on the bus—you don't have to throw them off. They can shout—you don't have to silence them. You acknowledge they're there, you notice what they're saying, and you keep steering toward your values anyway. The passengers might be noisy, but they can't actually grab the wheel unless you let them.

This isn't about suppressing emotions or pretending they don't exist. It's about recognizing that emotions are passengers, not drivers. They're part of the journey, but they don't determine the destination.

• • •

SHIFT 49

> **FROM:** *"This emotion proves something is true"*
>
> **TO:** *"This emotion signals something; it doesn't dictate reality"*

The distinction between signal and dictate is crucial.

Emotions signal things. Anxiety signals perceived threat. Sadness signals perceived loss. Anger signals perceived violation. These signals are worth noticing—they contain information about how your mind is interpreting your situation.

But signals aren't the same as facts. A smoke detector signals possible fire. It doesn't prove there's a fire. Sometimes there's actual fire; sometimes there's burnt toast. The appropriate response is to investigate, not to assume the signal is accurate.

Similarly, your anxiety signals that your mind perceives threat. The appropriate response is to examine whether there's actually a threat—not to assume the perception is correct. Your jealousy signals that your mind perceives some risk to the relationship. The appropriate response is to examine what's actually happening—not to act as if the jealousy is proof.

This shift preserves the value of emotions while limiting their authority. You don't dismiss them ("this feeling doesn't matter"), and you don't worship them ("this feeling must be true"). You treat them as information that deserves attention but also deserves scrutiny.

The practical impact is significant. When you feel certain something is true because it feels true, you're operating in emotional reasoning. When you recognize that the feeling is a signal worth investigating, you create space to actually look at the evidence. And often, the evidence tells a different story than the feeling suggested.

• • •

Cognitive Defusion Techniques

ACT offers specific techniques for creating distance between yourself and your thoughts and feelings—a process called *cognitive defusion*. The goal isn't to eliminate the thoughts and feelings but to change your relationship with them.

One technique is linguistic: adding "I'm having the thought that..." to the beginning of a troubling thought. Instead of "This is going to be a disaster," you say (internally): "I'm having the thought that this is going to be a disaster." The content is the same, but the framing changes. You're no longer reporting a fact; you're noticing a mental event.

This sounds trivial. It isn't. The addition of "I'm having the thought that..." reminds you that thoughts are mental events, not direct transmissions from reality. The thought might be accurate or it might be distorted—but either way, it's something your mind is producing, not something you're obligated to accept.

Another technique is visualization: the *leaves on a stream* exercise. You imagine sitting beside a gentle stream. Each thought that arises is placed on a leaf and watched as it floats away. The thought isn't fought or evaluated; it's simply observed and allowed to pass. This builds the capacity to notice thoughts without being captured by them.

A third technique is the *silly voice*: repeating a distressing thought in a cartoon character's voice—or singing it to the tune of "Happy Birthday." The content stays the same, but the emotional grip loosens. It's hard to take "You're going to fail" as seriously when you've just sung it like Mickey Mouse.

These techniques work because they disrupt the fusion between you and your thoughts. Normally, thoughts feel like you—like direct expressions of your identity and situation. Defusion reminds you that thoughts are passing mental events, like clouds moving across the sky. The sky isn't changed by the clouds, and you aren't determined by your thoughts.

• • •

The Stoic Pause

The Stoics understood this principle millennia before cognitive psychology formalized it. Epictetus taught his students what might be called the *discipline of assent*: the practice of pausing before accepting an impression as true.

"Be not swept off your feet by the vividness of the impression," Epictetus wrote, "but say, 'Impression, wait for me a little. Let me see what you are and what you represent. Let me test you.'"

This pause—this moment of testing before assenting—is the space between stimulus and response. The emotion arises. The interpretation forms. And before you accept it, you pause. You examine. You ask whether this interpretation is actually warranted.

Modern cognitive science describes this in terms of System 1 and System 2 thinking. System 1 is fast, automatic, and emotional—it generates the immediate reaction. System 2 is slow, deliberate, and analytical—it can evaluate whether the reaction is appropriate. Emotional reasoning happens when System 1 runs unchecked. The Stoic pause is the activation of System 2.

The pause doesn't guarantee a better response. Sometimes, after examination, you'll conclude your initial reaction was correct. But the pause ensures that your response is chosen rather than merely triggered. It's the difference between acting and being acted upon.

This practice is simple to describe and difficult to execute. When emotions are strong, the urge to act is strong. The pause feels unbearable—like holding your breath while everything in you demands that you inhale. But the capacity to pause grows with practice. Each time you insert space between trigger and response, you're building the neural pathways that make space easier to find next time.

• • •

What This Looks Like in Practice

Let's make this concrete.

You receive an email from your boss with the subject line "Quick chat needed." Immediately, you feel anxiety. Your mind generates the thought: "Something's wrong. I'm in trouble."

Option A (emotional reasoning): You treat the anxiety as evidence. You spend the next hour ruminating about what you might have done wrong. By the time the meeting happens, you're defensive and on edge. You interpret neutral comments as confirmation of your fears.

Option B (the pause): You notice the anxiety. You say to yourself: "I'm having the thought that something's wrong." You recognize this is a signal, not a fact. You examine the evidence: Has your boss given you any recent indication of dissatisfaction? Is it possible the "quick chat" is about something neutral or positive? You let the anxiety be present while investigating rather than assuming.

The anxiety might be correct—sometimes bosses do deliver bad news. But often, the "quick chat" is routine. Emotional reasoning prematurely commits you to a negative interpretation. The pause lets you respond to what's actually happening rather than to what you fear.

Or consider a relationship scenario. Your partner says something and your immediate feeling is hurt. Emotional reasoning says: "I feel hurt, therefore they intended to hurt me." The pause says: "I feel hurt—let me examine whether they intended this." Often, the hurt comes from an interpretation, not from hostile intent. The pause creates space to check the interpretation before acting on it.

• • •

Putting This Into Practice

TAKE ACTION: YOUR NEXT STEPS

Reflection Questions

1. Think of a recent time when you acted based on how you felt rather than on examination of the situation. What was the feeling? What did you assume it proved? How might things have gone differently if you'd paused?

2. What are the "passengers" that most frequently try to drive your bus? Which emotions or thoughts tend to dictate your direction when you let them?

3. When do you most struggle to create space between trigger and response? What conditions (stress, fatigue, specific people) make the pause hardest to find?

Action Steps

1. **Practice linguistic defusion.** For one day, notice when troubling thoughts arise and mentally prefix them with "I'm having the thought that..." Observe how this small addition changes your relationship to the thought.

2. **Implement the ten-second rule.** Before responding to any emotionally triggering message or situation, count to ten. Use the pause to ask: "Is my immediate interpretation the only possible interpretation? What would I think if I assumed positive intent?"

3. **Distinguish feeling from action.** When you notice a strong emotion, explicitly separate the two components: "The feeling is _____. The action I'm

considering is _____. Are there other possible actions?" Write this out if needed.

4. **Try the leaves on a stream.** Set aside five minutes. Sit quietly and notice thoughts as they arise. For each one, visualize placing it on a leaf and watching it float away. The goal isn't to stop thoughts but to practice observing them without being captured.

5. **Catch emotional reasoning in real-time.** Notice when you're using feelings as evidence ("I feel anxious, so this must be dangerous"; "I feel hurt, so you must have intended harm"). When you catch it, ask: "What's the actual evidence, independent of how I feel?"

Troubleshooting

"If I don't act on my emotions, aren't I suppressing them? Isn't that unhealthy?"

Suppression is pretending you don't feel what you feel. The shifts in this chapter ask you to feel what you feel—and then choose your response rather than being automatically driven by it. You're not suppressing the anxiety; you're feeling it and choosing not to let it dictate your behavior. That's the opposite of suppression.

"Sometimes emotions are right. Shouldn't I trust my gut?"

Sometimes emotions are right. The question is how you know which times. Trusting your gut unconditionally means trusting it when it's correct and when it's wrong. The alternative—checking your gut against evidence—lets you benefit from accurate emotions while catching the distortions. You're not dismissing emotions; you're treating them as one input among several.

"The pause sounds great in theory, but in the moment I can't think clearly. The reaction just happens."

The capacity to pause is a skill, and skills develop with practice. You probably won't pause successfully every time, especially at first. Start with lower-stakes situations where emotions are milder. Build the neural pathways when it's easier; they'll be more available when it's hard. Over time, the pause becomes more automatic—not less reactive, but more skilled at inserting space.

• • •

The final chapter of Part Four addresses not individual moments of reaction but longer-term patterns of resilience—how to sustain equanimity across the ups and downs of a life, and how to hold uncertainty without being paralyzed by it.

CHAPTER 15: Reframing Time and Change

The long-game perspective that makes everything more manageable

"Think of substance in its entirety, of which you have the smallest of shares; and of time in its entirety, of which a brief and momentary span has been assigned to you."

— Marcus Aurelius, *Meditations*

You're three months into something—a new habit, a career transition, a relationship repair—and it's not where you thought it would be. The progress feels slower than you imagined. The breakthrough you expected hasn't arrived. And somewhere in your mind, a familiar pressure builds: *This should be further along by now. Something's wrong. Maybe this isn't going to work.*

This is the moment where most change efforts die. Not from lack of ability or commitment, but from a mismatch between expectation and reality. We expect dramatic transformation on a timeline we invented. When it doesn't arrive, we conclude the effort is failing—and we quit.

But the problem isn't the progress. The problem is the frame.

• • •

Fortune-Telling and the Future

CBT identifies a cognitive distortion called *fortune-telling*: predicting negative outcomes without considering other possibilities. Instead of weighing probabilities or acknowledging uncertainty, we assume the worst will happen and treat that assumption as fact.

Fortune-telling sounds obviously irrational when stated plainly. But in the moment, it doesn't feel like prediction—it feels like seeing clearly. *I'm not going to get the job. This relationship is doomed. I'll never figure this out.* The certainty feels like realism. The anxiety feels like preparation.

The distortion works by narrowing the future to a single trajectory—usually the worst one. You don't consider that the interview might go well, that the relationship might improve, that understanding might come with time. You've already decided what will happen, and now you're living as if that outcome is inevitable.

Fortune-telling is particularly damaging during change efforts because change is inherently uncertain. You can't know how long it will take. You can't predict exactly how it will unfold. This uncertainty creates anxiety, and anxiety defaults to negative prediction. The more you care about the outcome, the more likely you are to convince yourself it won't work.

The antidote isn't positive prediction—that's just fortune-telling in the opposite direction. The antidote is epistemic humility: acknowledging that you don't know what will happen, that multiple outcomes are possible, and that your anxious certainty is a feeling, not a forecast.

• • •

All-or-Nothing Thinking About Outcomes

Fortune-telling often pairs with another distortion: *all-or-nothing thinking*. This is the tendency to see situations in extremes—complete success or total failure, perfect or worthless, solved or hopeless.

All-or-nothing thinking applied to change creates impossible standards. Either you've transformed completely or you've failed. Either you've reached the goal or the effort was wasted. There's no middle ground, no partial progress, no "better than before."

This binary creates a cruel dynamic. Because transformation is gradual, and because any given day might not look dramatically different from yesterday, all-or-nothing thinking consistently categorizes legitimate progress as failure. You're better at managing your anxiety than you were six months ago—but you still feel anxious sometimes, so you conclude nothing has changed. You're more present in your relationships than before—but you still get distracted, so you dismiss the improvement.

The distortion ignores that change operates on a spectrum. The question isn't "have I arrived?" but "am I moving?" Not "is this perfect?" but "is this better?" Progress measured in black and white will always look like failure because real change happens in gray.

When you combine fortune-telling ("this isn't going to work") with all-or-nothing thinking ("anything less than total transformation is failure"), you create a framework that makes sustainable change nearly impossible. You're simultaneously predicting doom and defining success in a way that can't be achieved.

• • •

SHIFT 50

FROM: *"I need this resolved now"*

TO: *"Some things take time; I can keep moving"*

This shift addresses both distortions simultaneously.

"I need this resolved now" contains a fortune-telling assumption: that if it's not resolved soon, something bad will happen. And it contains an all-or-nothing frame: either it's resolved or it's a problem. The urgency feels necessary, but it's usually manufactured by anxiety rather than demanded by reality.

The reframe—"Some things take time; I can keep moving"—makes two important acknowledgments.

First, it acknowledges that timelines aren't always within your control. Some problems are solved quickly. Others unfold over months or years. Relationships rebuild slowly. Skills develop incrementally. Healing happens on its own schedule, not yours. Demanding immediate resolution doesn't speed things up; it just converts ongoing progress into perceived failure.

Second, it separates resolution from functionality. You don't have to have everything figured out to keep living, working, and moving forward. Life isn't paused while you wait for the breakthrough. You can have unresolved situations and still act effectively. You can carry uncertainty and still make decisions.

This isn't passive resignation. "Some things take time" doesn't mean "nothing I do matters." It means you continue doing what you can while accepting that results unfold on their own timeline. You keep showing up. You keep practicing. You keep having the hard conversations. And you stop treating the absence of immediate breakthrough as evidence of failure.

• • •

The Stoic View from Above

The Stoics practiced a meditation called the *view from above*: imagining yourself rising higher and higher—above your room, your city, your country, the Earth—until you're looking at your life from a cosmic perspective.

Marcus Aurelius returned to this practice repeatedly in his *Meditations*: "Think of substance in its entirety, of which you have the smallest of shares; and of time in its entirety, of which a brief and momentary span has been assigned to you." The point wasn't to feel insignificant. The point was to gain proportion—to see your problems in the context of the vastness of space and time.

From above, the thing that feels urgent looks different. The crisis that seems all-consuming is a brief event in a brief life on a small planet. The failure that feels defining is one moment in a continuous flow. The timeline that feels too slow is a blink in the scale of things.

This isn't spiritual bypassing or minimizing real difficulty. The view from above doesn't eliminate your problems. It contextualizes them. It creates what psychologists call "psychological distance"—space between you and your immediate emotional reaction that allows for clearer thinking.

The practice is particularly useful when you're trapped in urgency. When everything feels like it has to happen now, zooming out reveals that most things don't. The project deadline matters, but it's not cosmic. The relationship tension is real, but it's one chapter in a longer story. Your growth trajectory is slower than you'd like, but you're comparing it to an invented timeline rather than to reality.

From above, you see that this moment is just this moment. It's not the final verdict. It's not the whole story. It's one frame in a longer film—and you don't yet know how the film ends.

• • •

Playing the Long Game

There's a reason most New Year's resolutions fail by February. They're built on a model of dramatic transformation: radical change, implemented immediately, producing visible results fast. This model appeals to our impatience but ignores how change actually works.

Sustainable change doesn't come from dramatic breakthroughs. It comes from small adjustments, maintained over time, compounding gradually. The person who loses weight and keeps it off didn't do it through a crash diet—they changed their relationship with eating, incrementally, over months and years. The person who built meaningful skills didn't have a flash of insight—they showed up and practiced, day after day, until competence accumulated.

James Clear, author of *Atomic Habits*, captures this principle: "You do not rise to the level of your goals. You fall to the level of your systems." The dramatic goal inspires the initial effort. The boring system produces the eventual result.

This is hard to accept because the dramatic breakthrough makes a better story. We want the moment of transformation, the turning point, the before-and-after. But real change rarely announces itself. It's happening while you're not paying attention—in the accumulation of small efforts that don't feel like progress until you look back.

Playing the long game means shifting your focus from the destination to the direction. Are you pointed the right way? Are you taking steps, however small? Then you're doing what you can. The arrival will happen in its own time—and demanding that it happen faster doesn't bring it closer.

• • •

Why Sustainable Beats Dramatic

The appeal of dramatic change is that it promises escape from the difficulty of the present. If you could just transform overnight, you wouldn't have to sit with the discomfort of being in process. You wouldn't have to tolerate the gap between where you are and where you want to be.

But dramatic change, when it happens at all, rarely lasts. The crash diet leads to regaining the weight. The radical commitment burns out into abandonment. The breakthrough insight fades without the infrastructure to sustain it. What willpower builds in a burst, willpower can't maintain.

Sustainable change works differently. It doesn't require superhuman effort. It requires showing up consistently, even when the effort is modest. It doesn't demand that you feel motivated every day. It asks that you do something anyway. It doesn't promise rapid transformation. It delivers gradual improvement that compounds over time.

Research in behavioral psychology supports this. BJ Fogg's "Tiny Habits" research shows that small, easy behaviors are more likely to become automatic than large, difficult ones. The person who commits to one push-up a day is more likely to still be exercising a year later than the person who commits to an intense daily workout. The small behavior creates a foothold. The foothold expands naturally.

The psychological advantage of sustainable change is that it doesn't require you to be a different person. You're not betting on a future version of yourself who will magically have more willpower, more time, more motivation. You're building from who you actually are, with the resources you actually have.

• • •

What the Long Game Requires

Playing the long game requires tolerating uncertainty. You can't know when the results will arrive. You have to keep moving without guarantees.

It requires patience with yourself. You'll have setbacks. You'll have days when you don't show up. You'll have stretches when progress seems invisible. The long game doesn't demand perfection—it demands persistence.

It requires letting go of comparison. Someone else's timeline is irrelevant to yours. The fact that they reached a milestone faster doesn't mean your pace is wrong. Different starting points, different circumstances, different trajectories.

And it requires trusting the process even when you can't see results. This is the hardest part. The absence of visible change doesn't mean nothing is happening. Skills are building beneath the surface. Neural pathways are forming. Patterns are shifting. The day it becomes visible is not the day it started—it's the day it accumulated enough to notice.

The view from above helps here. From above, you can see that this process has been unfolding longer than you realized and will continue beyond your current frustration. From above, you can see that your impatience is a feeling, not information about whether the change is working.

• • •

Putting This Into Practice

TAKE ACTION: YOUR NEXT STEPS

Reflection Questions

1. Where in your life are you engaged in fortune-telling—predicting a negative outcome with false certainty? What would it look like to acknowledge that you don't actually know how things will unfold?

2. What change efforts have you abandoned because they weren't producing dramatic results fast enough? Looking back, what might have happened if you'd stayed with the process longer?

3. What timeline have you invented for your own progress? Where did that timeline come from? Is it realistic, or is it an arbitrary expectation that sets you up to feel like you're failing?

Action Steps

1. **Practice the view from above.** When you feel trapped in urgency, take five minutes to zoom out. Imagine rising above your current situation—above your room, your building, your city, your country, the Earth. Notice how the urgent thing looks from this distance. Let the perspective shift your relationship with the problem.

2. **Replace timeline pressure with directional questions.** Instead of asking "Am I there yet?" ask "Am I moving?" Instead of "Is this resolved?" ask "Am I pointed in the right direction?" Progress is movement, not arrival.

3. **Identify one area where you can commit to sustainable effort.** Choose something you want to change, and define the smallest version of effort you could maintain indefinitely. Not the ambitious plan you'll abandon—the modest practice you can actually sustain.

4. **Catch fortune-telling in real-time.** When you notice yourself predicting negative outcomes, pause and ask: "Is this a fact or a fear? Do I actually know this will happen? What other outcomes are possible?"

5. **Measure progress in months, not days.** For anything that matters, zoom out when evaluating. Ask not "How did today go?" but "How does this month compare to three months ago?" The longer timescale reveals progress that daily measurement obscures.

Troubleshooting

"But sometimes things really are urgent. Not everything can be played as a long game."

True. Some situations require immediate action. But far fewer situations are genuinely urgent than anxiety suggests. The practice isn't to treat everything as non-urgent—it's to distinguish real urgency from manufactured urgency. When your nervous system is screaming that everything must be resolved immediately, that's usually a signal to question the urgency, not obey it.

"What if I've been patient for years and nothing has changed?"

If genuinely nothing has changed, that's worth examining. Are you actually doing something different, or just waiting? Patience without action isn't playing the long game—it's avoidance. The long game requires sustained effort, not passive hope. If you've been working at something for years without progress, it might be time to examine the approach, get outside feedback, or consider whether this particular goal is worth pursuing.

"How do I stay motivated when results are so far away?"

By not relying on motivation. Motivation is a feeling, and feelings fluctuate. Build systems that don't require motivation—habits that execute automatically, environments that make the right choice easier, accountability structures that keep you showing up even when you don't feel like it. Motivation gets you started. Systems keep you going.

• • •

This chapter—and this entire book—has been about one fundamental recognition: the frame determines the experience. When you change how you see your situation, you change your relationship with it. When you change your relationship with it, you change what's possible.

The fifty shifts we've explored aren't tricks or hacks. They're invitations to see more accurately. To stop treating feelings as facts. To stop predicting with false certainty. To stop requiring yourself to have answers you can't have yet. To stop measuring progress against invented timelines. To stop demanding resolution when you could keep moving instead.

You won't apply all fifty shifts perfectly. You won't remember them in every moment when they're relevant. But you don't need to. What you need is to recognize that the frame exists—that there's always another way to see this—and to get curious about what that other way might be.

The long game is the only game that matters. Play it.

CONCLUSION

Living the Switched Frame

"The happiness of your life depends upon the quality of your thoughts."

— Marcus Aurelius, *Meditations*

You've now encountered fifty ways to see differently.

Fifty invitations to step back from the automatic interpretation and ask: *What else could this mean? What am I assuming that might not be true? What would change if I held this more loosely?*

Some of these shifts probably resonated immediately. You read them and thought: *Yes. That's exactly the trap I fall into.* Others may have felt less relevant—problems you don't have, patterns you don't recognize. That's fine. This book isn't a checklist where every item must apply. It's a toolkit. You take what you need.

But here's what matters more than any individual shift: the underlying skill. The capacity to notice when a frame is causing unnecessary suffering and to consider whether another frame might serve you better. That meta-skill—the ability to examine your own thinking—is what transforms information into practice.

The question now is how to take what you've read and integrate it into the texture of your actual life.

• • •

Building a Daily Practice

Cognitive flexibility isn't a concept you understand once and then possess forever. It's a skill you practice. Like physical fitness, it atrophies without use and strengthens with consistent exercise.

The good news: the practice doesn't require hours of your day. It requires moments of attention, strategically placed. A few minutes in the morning. A few minutes in the evening. Brief pauses throughout the day when you notice yourself reacting strongly. That's enough.

The practice has three components: setting intention, catching patterns, and reviewing what happened. None of these is complicated. All of them require that you actually do them, which is harder than it sounds.

• • •

The Morning Frame Check

Before your day begins—before email, before the news, before the cascade of demands—take two minutes to set your cognitive intention.

Ask yourself: *What situation am I likely to face today that could trigger unhelpful thinking?* Maybe it's a difficult meeting. Maybe it's a conversation you've been avoiding. Maybe it's just the accumulated pressure of too many obligations. Identify it.

Then ask: *What frame am I likely to default to, and what frame would serve me better?*

If the difficult meeting triggers your perfectionism, remind yourself: *Progress over perfection. Done beats perfect.* If the conversation triggers your conflict avoidance, remind yourself: *Discomfort is not danger. I can tolerate this.* If the pressure triggers your all-or-nothing thinking, remind yourself: *I don't have to resolve everything today. I just have to keep moving.*

This isn't positive thinking. You're not telling yourself the day will go well or that difficult things will be easy. You're preparing your mind to meet difficulty with a more useful interpretation. You're loading the alternative frame so it's available when you need it.

The morning frame check takes two minutes. It costs you nothing. And it dramatically increases the odds that when the moment arrives, you'll have access to the shift instead of being swept away by the default.

• • •

The Evening Review

At the end of your day—before you collapse into whatever helps you decompress—take five minutes to review what happened through the lens of your frames.

Ask yourself: *When did I get emotionally activated today? What was I telling myself in that moment?*

Emotional activation is your signal. Strong reactions—frustration, anxiety, defensiveness, despair—almost always involve a frame. Something happened, you interpreted it, and the interpretation generated the emotion. The evening review is where you examine those interpretations.

Maybe you felt frustrated in that meeting because you were telling yourself: *This is a waste of my time. These people don't understand.* The frame was superiority mixed with powerlessness. What might have been a more useful interpretation? Perhaps: *This is slower than I'd like, but my job is to contribute what I can, not to control outcomes.*

Maybe you felt anxious about the email you received because you were fortune-telling: *This is going to be a disaster. I'm going to get blamed.* What did you actually know at the time? What were you assuming that might not have been true?

The evening review isn't about self-criticism. You're not looking back to judge yourself for having unhelpful thoughts. You're looking back to learn—to notice patterns, to recognize triggers, to prepare for next time. The goal is awareness, not perfection.

Over time, the evening review reveals your recurring themes. You start to see which frames show up most often, which situations trigger them, and which shifts you most need to practice. This self-knowledge is invaluable. You can't change what you don't see.

• • •

When You Slip (And You Will)

You will slip.

You will have days when every frame you've learned evaporates and you react from pure default. You will catastrophize. You will personalize. You will engage in all-or-nothing thinking about your own all-or-nothing thinking. You will wonder if you've learned anything at all.

This is not failure. This is the process.

The measure of progress isn't whether you slip—it's what happens after you slip. Do you use the slip as evidence that change is impossible? That's all-or-nothing thinking about your own growth. Or do you notice the slip, get curious about what triggered it, and return to the practice?

Slips are information. They show you where you're most vulnerable, which situations overwhelm your resources, and which patterns are most deeply ingrained. A slip doesn't erase your progress. It reveals where more progress is needed.

When you slip, avoid the second arrow. The first arrow is the original difficulty—the situation that triggered the unhelpful frame. The second arrow is what you do to yourself about it: *I should know better. I've read a whole book about this. What's wrong with me?* The second arrow turns a moment of difficulty into a cascade of self-attack.

Instead, try this: notice the slip, name the frame that took over, and ask one simple question: *What would I do differently next time?* Then move on. The slip happened. It's done. Your job is to learn from it and keep practicing, not to prosecute yourself for being human.

• • •

The Compound Effect of Small Cognitive Adjustments

Here's what you won't notice in the short term: the accumulation.

Each time you catch a frame and choose a different one, you're building something. You're strengthening neural pathways that make the alternative interpretation more accessible. You're creating precedents that your future self can draw on. You're proving to yourself that the automatic thought isn't the only option.

None of this is dramatic in the moment. You won't feel transformed on any given day. You'll just feel like you're practicing—sometimes successfully, sometimes not.

But over months, the compound effect becomes visible. Situations that used to send you spiraling now feel manageable. Reactions that used to be automatic now come with a moment of choice. You start catching frames earlier, before they've built up momentum. You start recognizing patterns before they fully develop.

This is what sustainable change looks like. Not a breakthrough that changes everything overnight, but a gradual accumulation of small adjustments that shifts the baseline over time. The person you are after a year of practice is different from the person you were when you started—not dramatically different, but meaningfully different. More flexible. More resilient. More capable of meeting difficulty without being overwhelmed by it.

Trust the accumulation even when you can't see it. The work you're doing matters, even when it doesn't feel like progress. The compound effect operates on its own timeline, not yours.

• • •

From Fifty Shifts to One Principle

If you remember nothing else from this book, remember this:

You are not at the mercy of your circumstances. You are at the choice point of your interpretation.

This is the principle beneath all fifty shifts. The situations you face are what they are. You can't always change them. But the meaning you make of those situations—the story you tell yourself about what's happening and what it means—that remains yours to shape.

This isn't about denying reality or forcing positivity. It's about recognizing that between event and response, there's an interpretation. That interpretation is often automatic, often invisible, and often unhelpful. But it's not fixed. With practice, you can catch it. With awareness, you can examine it. With skill, you can choose differently.

The fifty shifts in this book are specific applications of this principle to common situations. But once you grasp the principle, you can generate your own shifts. You can ask, in any difficult moment: *What am I telling myself about this? Is that interpretation accurate? Is it useful? What else could be true?*

The capacity to ask those questions is the real skill. The specific answers matter less than the asking. Because once you're asking, you're no longer at the mercy of the automatic. You're at the choice point. And at the choice point, you have options.

• • •

A Final Word on the Strategic Realist's Life

Throughout this book, I've tried to offer something different from both naive optimism and cynical resignation. I've called it strategic realism: seeing things clearly, accepting what you can't change, and focusing your energy where it actually matters.

The strategic realist doesn't pretend difficulty is easy or that effort always leads to success. But they also don't surrender to helplessness or assume the worst. They hold a middle position: honest about constraints, clear about what's within their control, committed to doing what they can with what they have.

This position is harder to maintain than either extreme. Naive optimism is comfortable—you don't have to face hard truths. Cynical resignation is also comfortable—you don't have to try. Strategic realism asks you to see clearly and engage fully anyway. It asks you to accept that some things won't go your way while still bringing your best effort. It asks you to hold hope and honesty simultaneously.

But this harder position is also more effective. The naive optimist is constantly blindsided by reality. The cynic never discovers what's possible because they never really try. The strategic realist navigates the world as it actually is, making progress where progress is possible, adjusting where adjustment is needed, and maintaining equanimity through the ups and downs.

The fifty shifts you've encountered are tools for this navigation. They won't eliminate difficulty from your life. They won't guarantee success. They won't make you immune to suffering. But they will change your relationship with difficulty, success, and suffering. They'll give you options where you thought you had none. They'll create space where you felt trapped. They'll help you respond instead of merely react.

That's what cognitive flexibility offers: not a better life handed to you, but a better capacity to work with the life you have.

• • •

The Invitation

This book ends, but the practice continues.

Tomorrow morning, before your day begins, take two minutes. Ask yourself what you're likely to face and what frame might serve you. Load the alternative interpretation so it's ready when you need it.

Tomorrow evening, before you're done for the day, take five minutes. Review what happened. Notice where you got activated. Examine the frames that showed up. Learn something you can use.

And throughout your days, in the moments that matter: pause. Notice what you're telling yourself. Ask if it's true. Ask if it's useful. Remember that you have a choice.

The frame determines the experience. And the frame is yours to switch.

UNLOCK THE SECRET VAULT

Scan the code below to enter the Vault and discover the missing piece of your transformation.

Unlock Your Exclusive Bonus Tools Here!!!

Made in United States
Orlando, FL
04 November 2026

90204191R00096